An Enthralling Thread

An Enthralling Thread

a student's journey

D<small>EEPAK</small> D<small>WIVEDI</small>

PARTRIDGE
A Penguin Random House Company

To order additional copies of this book, contact
Partridge India
000 800 10062 62
www.partridgepublishing.com/india
orders.india@partridgepublishing.com

CONTENTS

PREFACE

"What is the purpose of my life?" One day, this question knocked my mind. As always happens with questions, the question became of mine as it was. The question received my love and nourishment, and in the due course of time, it grew in a tree of questions; of several hundred branches, all slightly different in one way or the other. I had never seen such a tree ever, nor had I imagined. Fortunately, I soon began to realize that my questions were lonely, they were waiting for answers, and I, for my questions, began to search for the appropriate answers analyzing each of them for whether it qualifies my loving question or not. Whenever I found the most appropriate answer to my every question, I sang in emotion, and to embrace my songs, spurted the essays.

The answers belonged to my land, my only land, which the world is happy to keep dividing, initially in to four; the land of religion, the land of art, the land of philosophy and the land of science. I tried to walk as slowly as possible into the lands of the world, in order to stay longer. But I am someone with very little knowledge, and I could conclude this because, after wandering for 5 years, with a few answers, I finally had no answer to my first question; the more the facts I came across, the faster my conclusions collapsed. I strove to know how appropriate answers I have found for my loving questions. Under the effects of limitations, I began to preserve my songs and essays for I believed that they would guide me what better could be done.

"What has been my life yet?" overwhelmed by my curiosity, I turned my neck and with searching eyes, revisited my past without bothering that I was still stepping towards the future which might be more *uncertain* and *chaotic* than it was; I steadily focused on watching. The thread represents the sequence of universal facts. I have explored only a part of it with a few established universal facts and with the life events of my past 14 years including last 2 years of my stay with them. I was enthralled by the sensations the thread produced, these sensations took different forms in my vision. My vision is that I must reach you with this thread, because you are the part of it as is my momentary life which, as a matter of *chance*, became conscious of its own existence. I wish these enthralling sensations be replicated in you and quickly repaired if they produce errors, for I have learnt that we are the members of a family, *our land is one* and my palm small.

1

In the realm of thread

Someone has tied a thread to a dove's leg. The dove now looks better entertainer than ever. Although its inconvenience has grown, its adaptation skills have become advanced. The thread passes through the hands at regular intervals and some vocal expressions appear, "Hey! Don't fly the dove so high, you will lose the control and we are not sure of its landing."

I know this dove for two years. It has grown in my area. It is in its third year of survival and I, in my twenty ninth. We are similar in many respects. I see something like improvement in me too. There are three incisive reasons for this: First, working in Microbiology, I have recently completed Ph.D., and like most Ph.D. students in modern India, I own awesome experiences. Second, more recently, since my joining as Post doctoral fellow (PDF) in Indian Council of Medical Research (ICMR), my chaotic past has begun to fascinate me. The third and perhaps the most important, I am growing confident that I have learnt 'to observe' and that I can try to do a task of wonderful complexity—simplification.

India is a subcontinent of children where multiple constraints pervade the journey of education. Among those who manage to have a school, very few are provided with the occasions to realize the dignity of being a student. Twelve years ago in 1998, unlike my father, I did not need to struggle for education. My father, a doctor of medicine, never told me that he had been a meritorious student throughout his studies. I was little bothered about the exams being less concerned with the studies. I passed my Intermediate with 56 percent marks. I was happy. I had been focusing on many things since the exams were over. I was active in group discussions with classmates on quality education, top teachers, top cities in education, but now my major focus was on how father would react on looking at my mark sheet.

Father said, "I cannot predict where you will stand in future. Your path is long. Prepare well to become an ideal person. You don't know what it takes to deal with the limitations created by time. Now you do something that you have not done yet." and he placed the mark sheet into the record file.

My blunt intelligence had hardly met with father's words when his friend Dr. Shukla entered the house. He asked, "What about your result Deepak?"

"Yes uncle, I got it, I am pass," I replied.

"Show me your report card." I handed him my mark sheet.

"Oh! You have got grace marks in chemistry." his words were ear-piercing.

"Deepak you got only four marks in second paper. Are you not ashamed of such a performance?"

For this while my affection towards uncle and my hollow self-respect got entangled and shrunk somewhere in me.

I was feeling bad. I stood alone. The mark sheet was explained in two opposite ways; one associating with future possibilities and the other, linking with past doings. Relentless act of the thoughts filled up my eyes, I began to whimper.

"Don't cry my child, relax ... relax ... relax ... relax ... relax ... relax." Mother held my tear-saturated face in the bracket of her palms and slid the thumbs below the eyes.

Mother's heart has wonderful eyes; it foresees the emotional longings of the child. The numbered drops of child's sorrow tend to lose their content in the stream of mother's affection, and the reasons of sufferings lose their language.

My language has changed
Since I started breathing
I don't even remember the time
And prime concerns of dreaming
What was mine then?
O mother! I was you

A non-breather in a breather
The present bears the future
It does matter to live in a body
In a home living around somebody
What was mine then?
O mother! I was you

Something arrived something departed
Travelling through a thread
Before the consent, ahead of light
Your cells bestowed upon me 'the birthright'
What was mine then?
O mother! I was you

Had I not left you with pain?
"No!" you lie, don't you?
'I am' not in the world of many
I am a world inside you
What is mine then?
O mother! I was you

They say
They say, "Angles do come to take the great soul"
O, the lord of angles
I met you in my mother
They don't know
We live together
O mother!
O mother! I was you

My mother is a housewife. We are four siblings in the family. Among everyday responsibilities, mother always managed an academic duty while caressing me which was to say "Use your intellect in studies". This intellect was a big problem for me. "What the hell with it! Everybody says use it use it but how to use it?"

That day I perplexedly questioned to mother, "mummy, tell me once how to use intellect?"

"It is like I design a sweater for you," mother replied.

"You apply your intellect in it?" I asked.

"Yes. And when you say 'it is beautiful' your intellect is applied," mother replied.

"I can see the beauty of the sweater but not the intellect," I reasoned.

"Without the intellect it won't look beautiful," mother replied.

I got stuffs for a careful consideration. "Mummy, please apply the best of the intellect for my sweater" I broke the silence.

At the beginning of the year 1999, we shifted at Banda (UP). Since my fundamental knowledge was very poor, I realized that I must opt for B.Sc. in order to understand science well. Therefore, I did not even touch the idea of premedical examination coaching. I applied at Pt. Jawaharlal Nehru Degree College for the admission on merit basis. I was in waiting list. By chance, that year the college got the approval to open the Microbiology department. Luckily, I got the admission with a compulsion to have two additional subjects; Botany and chemistry.

Meanwhile, my sweater was ready. The sweater had greatly improved my confidence level. A fresh feeling of newness surrounded me everywhere I strolled in the college. Every student passing beside me seemed impressed by my new outfit. Eventually I got a new pinch from a tea boy at the main gate as he shouted in amusement "You have forgotten to zip your pants!"

Whatever had been used in making the sweater was practically important, and among the things left, was a spare yarn ball. I kept it with me as I could sense in it some potential for my future recreation. The city was new to me and I had no friends in neighbourhood and none at college either. At home my favourite pastime was to play with the yarn ball. The choice of the yarn ball had two distinct but related advantages; first, it was very handy from the safety point of view and second, it was noiseproof and I could have long playing sessions without disturbing others.

How enthralling the relation between the play and the player is! For a spectator, the play may be interesting or fascinating, nail biting or hair raising, effusive or boring, excellent or poor but the player never plays loaded with these adjectives. He is in search of that he knows as 'perfection'. For a player the play never loses charm, this is the reason why the great players are the best critic of their play.

The syllabus was issued by the department. At this beginning, I wanted to clear-up the blemish which my mark sheet had imprinted on my nascent self-respect. I thought of both the things, that the mark sheet can't be changed and that the disgrace of being an ignorant student can be removed. It was apparent in my mind that I am a poor student and that any sincere student would soon recognize this reality, therefore, I was looking for a fellow companion of my level.

As an important part of this whole plan-setup, it was necessary to get well-acquainted with the college in time. The college was not far from the house, and also, there was nothing wrong in reaching college with a purpose more than just study, my personal life was lacking the fragrance of college matters. All the necessary arrangements were being watchfully made; polished shoes, the shine in accordance with the hair, thoroughly rub-cleaned clothes and face, well-belted, two pens in the upper pocket, and a 6x9" executive notebook stylishly gripped between thumb and Index finger; a sincere student in outlook.

I had a charm for the college road. Though it was not solely the college road, it chiefly represented the college. The road had interesting elements: upward slope towards college; two speed breakers, some fifty meters apart, profoundly different in nature, one of them raised, made by municipality, and the other cratered, made by travellers; two types of bicyclists: cushionists (those who pass the upward slope in seated-riding mode), and situationists (those who pass the upward slope either in standing-riding mode or pedestrian-bicycle dragging mode); varied colours of youth-power; the persons going to college looked more focused than those returning. I, being unaware of this principle of path, did not wonder at my unfocused eagerness.

The house saw a revolutionary change. Landline telephone connection had taken place. This change made me far-reaching. Several times a day, in affection, I would grasp the receiver to listen the sweet music which I later knew as dial tone. On few occasions, placing the receiver back would result in the incidence of ringing. People seemed to be visible by ears. It appeared that the 'ear-telephone couple' was one of the most impressive pairs in this world. I fell in great admiration for the great inventor of this 'wonder-box' that redefined the world by distance. These days I was swiftly growing acquaintance in the city due to some dialling problems.

One day in the evening while returning from college, I was thrashed up by a boy of my colony. Actually, he asked me obstructing my way, "Oye! Where do you go with this hanging cover wear?"

"This is an apron and students have been told to wear it," I replied.

Then he began to sneer at me.

"You are stupid," I informed him and got blows.

I did not expose this matter in the house but the yarn ball got severe belting and lost its shape for my wrath. I stared at this tangle until I fell asleep.

In the realm of thread
I looked into the angles
Twists, turns and supercoils
Waves
Knots
I just kept on watching
Something

How wonderful it is! Thoughts recognize thoughts; they are the viewer and the picture, the listener and the speaker oblivious of their age and form.

Next day, on my way to college, I met another boy. "Are you in Microbiology department?" he asked.

"Yes," I replied.

"I saw you yesterday, nearby backside; why did he box at you?"

"Leave it. The details are bitter," I said.

"Yar, the same guy had once thrashed me too," he said.

This self-defined similarity made a stranger somewhat familiar.

I asked, "Your department?"

"Microbiology itself," he replied.

I had almost found a fellow companion although the similarity in intermediate marks was yet to be ensured. I asked, "What is your name?"

"Om," he replied.

"What was the percentage of your marks in intermediate?"

"75 percent"

"O my God! Twenty percent in surplus!" My plan was changed. I turned my face towards alongside grove and increased my pace.

2

I looked into the angles

Introduction, selection and involvement were pervading all over the college. A great deal of handshakes could be observed. The students were eager to obtain their identity cards. I anticipated that in some initial lectures, the students might have to reveal their previous marks, therefore, I did not pay any attention to the information about the class schedules. One afternoon, some students sprang towards me and began to explode one after another on my ears "We know you" "Often yow stroll with such a beautiful notebook all around the college" "Why don't you come into the classroom?" "Do you know your roll number?" I paid my sincere thanks to God for there was no one else except those whimsical guys.

My fear was taking colours of carelessness, and I realized it. I determined to attend the class regularly. The next day, I occupied the last seat before the lecture started. Upon entering the room, the teacher said, "Some students are not aware of the timings, the roll will be called right after the lecture." and started his lecture. Meanwhile, I got into a sudden contest. Fear confined, mosquitoes aligned: a man, skin tan. I could not tolerate those progressive bites and stood up.

"What's the problem?" asked the teacher.

"Sir, mosquitoes," I replied.

"What's your name?"

"Deepak Kumar Dwivedi"

"You seem to have come to the class for the first time. Is this the reason for not attending the class? You better join a mosquito awareness course." then there was laughter all around, and subsequently, I started taking a mosquito repellent coil in my pocket.

My thrilling trips to college made me sure of the fact that two types of studies were being conducted in the college. One, for which the admissions are granted and which depends upon the teacher, and the other, from which the teacher is generally excluded, which he is aware of, yet chiefly unconcerned until a fundamental mistake is made. For a boy, this study searches something in a girl and the other way round although the approaches are profoundly different on each side.

The boy side was reverberating with electrifying experiences. Most of the boys were awfully sure of the acquiescence of the other side. 'Misunderstanding' was a baseless assumption although few had a feeling of substandard possibilities with different outcomes. 'Soon to be thrashed' list was prepared by the powerful candidates and the enlisted ones were rambling obliviously with the future beaters. Appropriate points for different purposes were being searched. Updated versions of better abuses were released, three main patterns of their uses were apparent: two or three words in every third or fourth statement, the contents of these words being the body parts or body activities; one or two words in every sixth or seventh statement, the contents of these words being the family members; one word in every ninth or tenth statement, the contents of these words being the animal species.

I could not conclude anything as concrete about the girl side. I found myself drifting between reality and mere appearance whenever I looked at that side. I realized that for a boy this task ultimately seeks the help from girls themselves, and it was not necessary for me to go for that. Yet, being an ordinary observer of the dispersed events I found enough reasons to wonder at their mysterious activities; to hide as naturally as the Earth in its galaxy that the main focus is actually on the boy reaching nearby. Glitters of delicacy in the pace even if they are in all probability of jostling the female companion walking along. Tactful disposition of provisions. Keen and safe use of information technology. A boy of the class was fighting with somebody near college at 10 AM and before he appeared in the class at 10:10, the girls had all the discussion over, two of them made a late entry in the class, they gathered all the information regarding the matter by eyebrow and lip reading while taking their seats.

Few days later, in Microbiology period, the teacher put a fundamental question to the students, "what is the central dogma of genetic information?" Every student had to answer. Unfortunately, my turn came too early and I, without any preconditioned assembling of words, blurted out, "The threads of life are deciphered in the cell. The languages and the meanings travel through DNA, RNA and Proteins. The processes of formation of these threads are called—replication, transcription and translation, respectively." To my surprise, the answer was appreciated by the teacher and for the first time I found sincere students praising me. My shy nature and respectful manners put me in good books of the teachers and to nurture my self-respect via meeting their expectations, I initiated to live like a sincere student.

Om and I lived in adjoining colonies. He took me as a studious boy and began to commend the way of my imagination; however, I had no such conception about me, but I was sure about him that he was far ahead of me in correlating the things. Chemistry was his favourite subject. Soon, he was declared as the best student ever in the history of the college by the chemistry teacher while I was the favourite student of our Microbiology teacher. We established a good friendship and became the two never separating students of the college.

My attitude was changing. Attitude is the live signature on the mind's portrait: The assemblage of repeated sensations; Construction of representations along the inclination; delineation of a dance in a movement; attitude specifies mind's assets.

Few students in the class, from a recreational viewpoint, preferred me to Om, but, among us was a boy whom I liked for his unique biomechanical approaches: to cram the books in the shirt; to shake hands with a characteristic shrug and a radial smile; too fond of arm wrestling to prevent himself playing even with a child; almost in tears while seeing me off at my house; to hold the bicycle handle single-handedly in its middle, paddling it with full vigour upon seeing an approaching girl. He was in one-sided love with a girl who used to chew betel nut and therefore, he too.

In the month of May, Indian border witnessed the infiltration of the neighbouring residents. People were dying. I was focused on the daily updates of bloodshed. Bloody business! Is this what the tradition has given you? Or you have learned it? How has the globe become so complicated? Does humanity need such intelligence? Is this the philosophy of Aristotle? The Yoga of Patanjali? Is this the core concern of Galileo, Newton, Nobel, Einstein's discoveries? Is this the Darwin's natural selection? Are these questions orphaned? Have no validity? So, no answers! Or they are hidden? Human progenies, oh!

I see the feet
Who jump, trudge and come through
I see the waist
Belted, laden and straight
I see the hands
Who haul, clench and are injured
I see the chest
Playing with cannon balls
I see the eyes
Who lose the bisects
For a mission
O warrior!
You have to fight
Until a human awakes
My obeisance!

Struggle shadows the life from its origin. All life forms be it a primitive brainless cell or a multicellular intelligent organism, struggle for existence. We learn the aspects of life but we often forget that we possess something which helps design new means of war.

When an innocent child plays with his own shadow, he is unaware of the separation which has its bright side and the dark side as well. Through some awareness he finds that the motion plays for shadow sizes and as he recognizes the reality, he determines to restrict this darkness for a path.

Indian army launched Operation Vijay (Operation Victory) in Kargil. Soon after the Kargil Vijay Divas (Kargil Victory Day; July 26), B.Sc. part one result was declared. I secured 67 percent marks while Om obtained 76 percent. We both were contented with our performances and had no feeling of competition with each other. With the commencement of new academic session, we got indulged in studying together. Since my house was at an easy location, this was the preferred place of study. My parents were delighted to see my attention to studies. Mother always devoted great care to arrange us snacks and father would himself accomplish my must do household tasks. Mother would take care of the book for which I said "I like this book". I grew interest in book pages while mother, in book cover pages.

Like most other families in India, I grew in a religious atmosphere. My paternal grandfather is a priest. Maternal grandmother had lost her husband at my mother's age of three. The day onwards she gave up sleeping on the bed and lived in two ways: Spending days absorbed in Samadhi-like state, being unaware of bodily sensations, and crying out for Lord Krishna. She lived in village and I got to visit her once in every two or three years. I lovingly dared to touch her world of experiences with my pointed questions; she revealed her experiences recognizing my curiosity. It took me over a decade of investigations to understand her answers to my some of the earliest questions.

My question: Grandma, what makes you delighted and crying taking Lord Krishna?

Her answer: His play.

My question: Are there times when you forget about the family, relations and everything that surrounds you including the feel or sensation that you are sitting or lying down?

Her answer: Yes my child. As soon as Krishna disappears, I find that I am confined in a women's body and all the feelings related to the body crop up.

My question: How does Krishna look like? How does he speak?

Her answer: Krishna never looks same all the time. He changes his face, colour, voice, body, sometimes many handed with many faces, differently clothed everytime.

My question: Then how do you recognize him?

Her answer: When I call him "Krishna . . . Krishna . . . Where are you?" He comes and says "Lo! I have come, why do you worry?" Sometimes I refuse to accept him saying "You are fooling me, you are somebody else, not my Krishna." then he laughs and says "Look at me, am I not your Krishna?" then I say, "O, yes yes you are my Krishna, where had you gone leaving me alone?"

My question: Does it happen everytime? What else does happen?

Her answer: No, not everytime but most of the times. One thing that always happens just before he comes is I feel that my whole body energy is uplifting, intense painful feeling occurs, very painful as if somebody is dragging whole power of my body upwards. I cry "Krishna . . . Krishna . . . Where are you? Where have you gone leaving me alone?" then he comes and says "Why do you worry, nothing can harm you, you are mine, I am yours, this is all my Maya, I play for you and will continue to play, wouldn't you let me play?" then I say, "O my Krishna, play, play".

My question:	Does something that you only hear but cannot see and the other way around happen?
Her answer:	Yes, There are sounds, "Don't do this, it is dreadful". "She is my dear". "I will punish him who troubles her". "It is not where life ends". "You have to bear the pains". "You have to live". I often ask, "who is there? Krishna? Why are you not coming before me?" but nobody replies and a pin drop silence occurs. Then, there are some unusual faces having very big eyes, very big ears, silently gazing at me, they never come close, they never reply, they just gaze at me and disappear when I ask, "Are you Krishna?"
My question:	Do these sounds and faces appear when you eat or talk with family members or guests?
Her answer:	Yes my child, but not always. They disappear when I ask, "Who is there?

Krishna? Sometimes our relatives find it awkward.

My question: Do you have any body pains? Any injury?

Her answer: No my child, No injury as far as I can remember, but sometimes I have fever and minor body pains. Once a cow was reported running towards me, people shouted, "O save her, she will die, she is very old." but the cow was actually not running towards me, she was coming towards me, she came and sat near me. My only body pain is when my power is being dragged upward, that is very painful, I often complain to Krishna but he always replies, "Why do you worry? I am yours".

My question: How did Krishna begin to come to you?

Her answer: My child, your grandpa was the landlord of this village. He preferred services and joined police department. He was very intelligent and health conscious. One day when he returned from his morning walk, he said, "I have travelled through many forests walking alone day and night with no fear of death, today I saw a stranger in our field who stared at me in such a way that I felt I am going to die." Few days later he said, "I am having headache, Ah! See, Lord Krishna is here." and passed away. Bitti (My mother's nick name) was some 3 or 4 of age. My child, I am illiterate, I knew no special method of worshiping, no mantra, I would just light a lamp and sing few lines of a devotional song in Hindi, I learned somehow.

Some three years later, one day I felt that Lord Krishna was not there with me, filled with deep sorrow, I began to cry calling "Krishna . . . Krishna . . . Where are you? Where have you gone leaving me alone?" Bitti came and held me saying "Mother, O my mother what happened? Who has troubled my mother? Who is it? Mother, my mother, what happened? Tell me who has troubled you?" I said, "My child, Krishna has left me alone, I am missing him." "You call him he will come, he will surely come on your call, but call slowly and don't cry, You are a daughter-in-law of this village, what people will say? They will make fun of you." said Bitti. I wondered how Bitti had such a sense of understanding at the age of seven!

Next day, in the afternoon, the same thing happened and Bitti said, "Mother, you call Krishna, he will surely come, but don't cry, I can't bear people laughing at you." "Ok, I will not cry, but, I want nobody disturb me during all this, therefore I will sit in the innermost corner of the house," I said. "Ok mother, you sit there and call Krishna," said Bitti. I took that place and began to call Krishna, then began to whimper remembering what Bitti had said, but I found it impossible to control my cries, painful uplift of power happened for the first time and I fainted.

Few hours later, I found my head in the lap of Bitti sobbing. "O my Bitti, Don't sob, I am fine, I was sleeping," I said and clasped Bitti. "Mother, never leave me alone." Bitti griped me. "Never, never, my child, but don't get worried while I weep or sleep," I said "Ok mother I won't take worries," said Bitti. After few days, I thought may be Krishna got sulked because I don't go temple, so I started to go to a nearby temple but could not stop murmuring 'Krishna . . . Krishna . . . Where are you? Where have you gone leaving me alone?' People began to notice and wonder at my situation.

Some 20 days later, I fell ill, unable to move, I called Krishna day and night until I fell asleep 'Krishna . . . O my Krishna . . . I am unable to come temple, I am unable to come, what to do, O Krishna . . . O Krishna'. This thing continued up to some two weeks. Then, one day in the afternoon, that painful uplift of power happened and I heard a voice "See, I have come, don't come temple, you just call me, I will come to you with all the temples." "Who is it? Come forth, you are playing with me?" I thundered, but his laughter became louder and louder, he said, "Yes, this is my play and I will continue to play, you don't remember I am playing with you for long, who are you? You are mine, you have to bear your pains, this is the part of our play, I will come on your calls, I am yours, look at me, am I not your Krishna?" "O Krishna, O Krishna, My Krishna, where had you gone leaving me alone?" I complained, wept and then we both burst into laughter.

My question: Have you seen any ghost?

Her answer: I see many strange individuals; they don't look like humans or animals, and then I ask, "Who is it? Krishna?" no reply occurs if Krishna is not there, what to do with ghosts etc?

My question: But, some people including our family members say that once you communicated with a ghost of a person named Sannaam!

Her answer: I didn't know whether he was a ghost! He did not tell me about this, nor did I ask him. What actually happened is many years ago when I was returning from a temple which is few kilometres away from our village, it started raining heavily, for many hours, and I got confined at a place. The water flow took my slippers away; nobody was there to help me out, I called Krishna for help, a voice echoed 'Sannaam'. I shouted, "Sannaam!" A person came and said, "Yes". "Bring my slippers." I ordered. He brought them and helped me to reach the village and then returned away. People asked me,

	"How did you get here in this heavy rain? Who has brought you here?" "Sannaam," I replied. "Who Sannaam? There is no Sannaam in this or neighbouring villages." they wondered. Later they identified that there lived a man named Sannaam in the neighbouring village who died several years ago.
My question:	Grandma, why this universe is so complicated? Why do we wander all along the life? What is to be achieved?
Her answer:	My child, I cannot understand your questions, what do I know about the universe and its relation with life and its achievements? I am illiterate.

It is since March 23, 2012 that I cannot locate her in physical world which we know as ours. She was the most innocent devotee I have ever seen. For my family, she was the biggest example of a simple life full of explanations for several issues of our complicated world.

One afternoon, when I was passing through the court's backyard, a priest, who was sitting on the terrace surrounding a Pipal tree, called me, "O child! Will you please help me write the name of Lord Rama? I am too old to do it for so long."

I greeted him and asked, "Ok, where to write?"

"On this tree trunk, take this paint brush."

Following his all the instructions, I completed the work.

"You are a good boy. I wish you learn astrology. Will you do this?"

"Yes Guru Ji," I replied

"Ok. Come from tomorrow," said Guru Ji.

Often we are bewildered to see the happenings in our lives. We feel confused, failing to comprehend the pattern which seems unusually different from our known momentum. We see ourselves moving into a hazy picture. We humans are watchful beings; we carry the things of past into present, someway, and this is the way mere possibilities define themselves. We uncover a conditional possibility—acute angles merge into a straight line.

The addition of this new subject to my studies tightened my daily routine. Moreover, I had to study the literature of ancient Indian astrology in Sanskrit, the mother language of many modern languages. I discovered that I am not so poor in Sanskrit and to learn the verses was not a tough task. I took little interest in calculation part but developed a keen interest in interpretation of chakras. Guru Ji always appreciated my style of interpretations.

In search of the meanings, we begin with a limited set of information and progress towards the infinite. Our comprehension of languages tends us to reason for how a dumb speaks, a deaf hears, a blind sees, how an agonized one sings and a happy one stays silent. We embark upon a journey that rises as a source, moves as a message, sustains as a channel and signifies as a destination.

During this transition period, my joint studies with Om became less frequent. The exams were nearing. Our medium of study was Hindi and we had to translate the literature from English to Hindi. We had prepared the notes but did not distribute them between us considering that we would do so according to the requirements. During exams, I was missing few important notes but Om was not interested in searching them. The notes were not retrieved and it was impossible to prepare them again. Studying alone, wet eyes, hazy sentences, dot like words and no letters. Mother would console leaving all her works aside often saying "only three things are yours: your skill, your destination and your path."

He peers at a dream
The lonely dream
He peers at a dream
There is something in his eyes
The doubts are on the dream
At the tip of the real highs
In the haze
Preceding the future
The dream speaks
To the explorer
O Seer!
See yonder

The examinations were over. Talks ceased between Om and me, and Guru Ji left for a pilgrimage. Meanwhile, another classmate and Om's neighbour, Shailendra became a close friend of mine. I could clearly see two things in him; a growing interest in studies, and a keen attention to fights; highly aware of invariably reaching the market to clear the fight accounts and to evaluate further options.

On one occasion in the market when a dispute was cropping up, he told me, "Deepak I want you go home right now, keep away from the quarrels." I said "No, I won't, what are you going to do here? If this is your matter, I will face it." Then he said "Ok, we both are going home." He dropped me home and returned back to take part in the fight.

Later on, when I got the news of the matter, I asked him, "Now that your all time fighting friends seem to be more important than your studies and my friendship as well, you better leave my company. I suggest there is no need to assemble more and more incompatible things." He promised not to fight again.

B.Sc. second year result was out. In spite of having such an abruption during the exams, I was sure of not falling short of my previous performance levels. I got exact 67 percent marks again. Om also maintained his previous rhythm. Om did not reveal his marks to me nor did I ask him for anything. The teachers and the students would often say, "You both don't look together these days! It seems something lacking; you have been a fresh example of behavioural synergy between the students."

I got driven towards a property of the thread—flexibility. Sometimes, a simple progression transforms into a strange hanging pattern. The eyes won't focus on this change; they want their own sights but the intellect doesn't approve this demand. In the space behind the eyes, between the ears, uprising thoughts soar higher and higher. In the absence of the destination, origin pulls them and they then, fall embodying heaviness; a thought never shatters, it is flexible enough to acquire a new form. A new thought, thus formed, often gets an unconditional support from an innocent hope.

3

Twists, turns and supercoils

In the third and last year of our graduation, Microbiology department was having some important changes. The fund for the department was sanctioned. The progress could be clearly seen on the walls of new building, on the tables equipped with drawers, on the new instruments tidily arranged in the laboratory, and on the faces of students confidently sitting on new chairs. I heard two teachers talking about the upcoming events "Now the students will get a better atmosphere for learning, trainings on mushroom cultivation and food preservation will be conducted, lecturers from other universities will be invited every fortnight for guest lecture. A new Microbiology teacher has also been appointed; he will soon join the department."

The principle of progress of the college and of the student works on a simple relation; the progress of the college depends on the progress of its student. You just walk through any progressive college of the world; you will see how the glory of the student takes you to the pride of a teacher and to a great foundation known as college.

In this changing atmosphere, my biomechanically unique class fellow, who was chewing betel nut falling in one-sided love with the girl that chewed betel nut, one day, in the presence of the girl exclaimed, "How joyous it is to chew the betel nut well mixed with tobacco! I enjoyed the last weekend with it; this week is also going excellent!"

Suppurating emotions were unveiled. I was stunned. I have been taught about the downfall caused by intoxicants during my childhood education but on this occasion it became extremely tough to me to define 'downfall'. Is it about physiological fitness or moral fitness? Both? I saw my class fellow in a pitfall on the road to morality and it was so because I knew much less about physiology as compared to moral awareness. Those days the slogans of joy on tobacco product packs did not share the fact which is today put as statutory warning concerning cancer.

I gestured him to meet me separately, he trudged towards me. "Have you really started to take tobacco?" I asked.

"Yes," he replied.

"Why?"

"Because she has begun to have it."

"What? Are you sure?" I wondered.

"Yes, this is well confirmed, I saw her purchasing it, I knew the brand from the shopkeeper, observed her spittle and confirmed by comparing it with my spittle," he confidently replied.

"But why are you getting into such bad habits?" I urged.

"I don't know," he muttered, and rubbing the tobacco spot of his palm, he left with a drooping head.

I attended his words roving in my mind for long, and now I could clearly see him unfolding a truth of their joint history and sending words to her for which my presence was crucial. Consequently after few minutes, I found myself with a superset of twisted thoughts wherein the most featured were two questions: would his recent effort get the advantage of the fact that intoxication pulls two consumers closer? Or would he be able to put the foundation of a true relation by putting one into the realization of one's mistakes? Alas! The truth my classmate's innocent heart soon met with was bitterly surprising; the girl was already in deep love with a man several years older than my friend.

I got to know that our new Microbiology teacher Mr. Rajeev Jain has come, and for the time being, is accommodated in the college. Few students had even met him twice; I was waiting for his first lecture and was not interested in any private meeting with him. On his first day in the class, we found him seated, waiting for us. He got so absorbed in the introduction that it looked impossible to get introduced with all 32 students in his first lecture spanning one hour, but he got well introduced with all the students. I was able to understand that he was teaching something by studying the students. The introduction was an eye opener to me as it surprisingly revealed the status of preparation of several sincere students.

The teacher and the student: the blending of colours, colours of efforts; of search; of faith; of success; a portrait in making; a sight of Intellect thriving in the cascade of reasons.

Have you ever seen a student wandering in search of a book? Books are the testimonies that illustrate the relationships, and at heart is a joint enterprise that imparts the gravity and depth to the book and to the teacher, and the heights to the student. In India, the student in his primary education learns a bow: to touch the feet of the teacher and to touch the book with his forehead.

They are never small or big, self or non-self
Either they are or they aren't
The book, the teacher, the student

Mr Jain completed the first month of his teaching. Now, there became two teachers of Microbiology in the Microbiology department so the teachers of botany and chemistry were relieved from their extra-duty of teaching Microbiology. Students were happy as the things looked easier to understand. But, slowly in the course of past two years, a trait became protuberant on the studentship of the students: pungent competition; rich in diplomatic craft.

I had no idea of such tricks. Om's friendship had not let me move my head to this scenario; I caught the only glimpse of his competition during previous examination. But here had been assembly works, coalitions, separations, reunions; everyone conscious of swooping on the best literature and then concealing it. Pervading motives had firmly tied the study subjects to the students. Distance from Om was parching my studies; I had learnt to study from him. This year, I was having a session full of secluded pangs.

It is very interesting to a student that his enquiry with his little knowledge takes him to the big question. Once he reaches to the relevant question, the shadows of answers appear. The sphere of answers around the question becomes smaller and smaller, and then he finds a principle with one answer. The question and answer construct a single fact; they have different bodies, one soul. Are you in search of fundamental facts? Go, read the forehead of a student, where, in the contraction full of questions and in the expansion full of answers travel the facts, this is the secret of student's shining forehead.

By the second month of his teaching Mr. Jain became dear to all in the college, always surrounded by the students. Seeing his fond efforts for solving the problems of the students, I often questioned to myself, "Do I love to study as much as my teacher loves to teach?" I could never afford an answer. Slowly, I fell into possession with this unanswered question, and it became necessary to keep on to the right path with the right pace. The task didn't look so difficult in Mr. Jain's presence.

During Mr. Jain's lectures, Om often appeared as if he was not a single student, instead a group of students; putting multiple questions of various objectives to Mr. Jain. Few of his objectives were well known to me, so even knowing that he never assembled a question before any thorough preparation, I would try to answer him. The questioning-answering that had been once a part of our private joint studies now became the part of our class studies. To answer Om's questions ably was the only parameter of my knowledge in my eyes. This perceptual mistake placed me closer to Mr. Jain and I began to observe his hopes. Coincidently, a classical atmosphere for learning was developed in the class; Mr. Jain initiated to extend Om's questions class-wide, waiting for the relevant answers to appear. The attestation of Mr. Jain's curiosity led the students to cognize the fragrance of their development.

I had no special idea of my individual qualities, no idea that my presence may claim such an importance to Mr. Jain. One day after the class, while strolling with me along the college playground, he took out a box full of sweets from his bag, and said, "You are not of Om's type, you both are of different types; profoundly different. Last year you obtained nine percent less marks than Om, I bet this year you will secure at least nine percent more than that you have obtained, *Rasgulla* (a famous Indian sweet which is prepared from milk products) is my favourite sweet, if I lose, I will never take it." and offering me two *Rasgullas*, he briskly finished the rest.

After his departure, I left for home with a new speed. But, after a few minutes walk, I found myself hanging on the question marks of varying sizes, drifting in the air flowing from two opposite sides. "What 'type' of student am I? Is it not like that I have been overestimated by the dignity of Mr. Jain's vision? How would I gather those marks of merit?" The *Rasgullas* whom he wagered, and whom he was going to crave for in near future, were now appearing to me hanging on the trees of the college road.

The fourth month of Mr. Jain's teaching was full of practical demonstrations. Now he was on the least vocal use of words. His serious silence was getting louder day by day. He was collecting books, studying much more, repeatedly saying "The days of hard work have begun". I fell in to serious doubts "whose days of hard work have begun?" "I have to appear for the yearly exams and Mr. Jain is studying as if it were his exams!" With all my silent queries, I kept noticing my teacher growing into a student.

One day he called me up and said, "I had come here with an only purpose of teaching, now I am focusing on my next aim."

"What is that, sir?" I asked.

"I am in love and much is left to deal with for our future."

"My God! Then it is very necessary to find out what she likes to chew lest the same thing drag Mr. Jain into the situation as my biomechanically unique class fellow got into" The two connected thoughts began to roll with different speeds in my mind.

"May I know her name? Has she been informed about all this?" I asked.

"Anamika Jain (*Sanskrit*; the word Anamika means *she who has no name,* also denotes *the ring finger*). She also loves me; she will talk to her parents regarding our marriage but she wants to see me qualifying NET (National Eligibility Test) so that parents will be having an assurance of our well settlement and our relation will not be frown upon taking any such issues. I have tried to put my best all over but if I continue to stay here and meet Anamika frequently, I will surely be unable to fulfil her wish and many lives will be destroyed. I need to concentrate well therefore, I have tendered my resignation; leaving for my hometown," he replied.

It sounded to me like I was on a hammock hanging on the clouds; fully uncertain of how much time would it take for a complete oscillation to occur. The thread, from which I had learnt the principle of flexibility, was now waving before me embracing the forms of twists, turns and supercoils.

January 26, 2001. The republic day celebrations of India were eclipsed by the cries. The cries were coming from Gujarat; so many cries: sunken cries; tattered cries; dumped cries; two-footed; many-footed; feathered; orphaned; earthquake; earthquake; earthquake.

Lap is snatched away
At a doorway
The child screams

Dreams are sawn
Trampled on
The adolescent screams
Clan is demolished
Supports are banished
The aged one screams
Clamour environs
Danger beacons
The beast screams
Entrapped between ruin and fortune
The life screams

The earthquake reached a magnitude of between 7.6 and 7.7 and killed around 20,000 people, injured another 1, 67,000 and destroyed nearly 4,000,000 homes in two minutes. When one eye beholds the destruction and the other becomes the part of it, one learns how far the two eyes lie.

In the city of grievances
On the wounded paths
Piercing the air
Changing the volume
And burying the colours
Rays carry away the sorrow drops
The disaster stops.

Mr. Jain headed towards his destination. Entire February, he had been divided into two parts; Anamika and me. Seeing him off, we embarked upon our separate journeys with our respective Mr. Jain. His affection had made my aim very personal to me and my path, enthralling; his affection and my aim were not two separate things. Anamika had to set to a knotty enterprise;

wait of her love's success; how would her aim be explained in a straight way? Every day, I would go to the bed complacent with my progress, remembering Mr. Jain before falling asleep, while she would extract tears from those hauntingly beautiful memories and dangerously preoccupying consequences. Sitting on the study chair, I would share my daily achievements with my parents any time in the evening, while she would cast herself here and there in the house concealing her situation all day long. The aim that Mr. Jain had left with was in fact a set of two types of pursuits; that I was *moving* with and that Anamika was *staying* with.

Among the most uproarious phenomena that I have seen are 'life', 'struggle', 'peace' and 'love'. In various definitions of love, you may find a detailed description of the importance of the things that pertain to it. In many of these, you will find that it is the 'beautification' of emotional expressions that relate to belief and feelings. Many put it as a 'social interpretation' that brain makes out of the physiological processes in the body. Many other imply that it is a 'cocktail of reality and imagination', and several others conclude it as 'the heart of self-contradictory experiences'.

Perhaps I will never reach at anything as 'complete explanation of love', yet, so far as I can reason, no isolated existence of love is possible. Various existing interpretations of love are of great importance as they are the efforts of great thinkers with a quest for an ultimate explanation. Whatever I have seen happening in love has led me to grasp something with my little knowledge: *whom love touches for once, becomes an eyewitness to the grand picture; imaginations on one side, evidences on the other; on one side are the gatherings of loneliness, on the other is the solitude of omnipresence; ahead is attainment, renunciation behind.*

Have you been searching where love grows? In a social and geographical sense, one may seek it in the lands where humans and animals have 'sharing' and in a biological sense, in the organ of animal origin—brain. *Love alleviates the pain of limits, it is a natural soother, manifests through attention, sways in the cradle of emotions, plays in feelings, adorns in thoughts. It belongs to the dweller in the body; a mother would soonest recognize this truth. We love life, we live love, we live in parts, we know love in parts, to live love in its totality is the ultimate goal of humanity.*

In the first week of July, the final result of our graduation was declared. The first glimpse of report card made me feel like I was on the wing, as if, in a moment, having soared high in the sky of past, I had landed on the earth of present. How powerful the seed of Mr. Jain's true bet was! It had given rise to everything in time; root of realization, stem of steep efforts, leaves of lucid comprehension, blossoms of achievement with a crown of blessings. Mr. Jain was right; I had obtained 76 percent marks.

Happiness is so amazing; it runs backward haphazardly tucking the edges. I was yearning to meet Mr. Jain but I had no information of him, no address, no telephone number. I wished somebody tell him that his *Rasgullas* were now out of danger. For three days I rambled in the carnival of thoughts. On the weekend, father told me to fill up the M. Sc. Entrance examination form of Bundelkhand University, Jhansi, I got focused on preparations.

Om qualified the test and I too. I took admission in M. Sc. Microbiology but Om didn't. He headed towards a thorough preparation for Civil service examination and left

45

for Allahabad (UP), the hub of students, especially those aiming to be among the future IAS and PCS officers. I was leaving home for the first time. Mother had stitched worries on her forehead. At the station, few minutes before the arrival of the train father said, "It is usual to know where one has to begin from but rare to judge at the beginning where one will reach."

In due course of time, I realized that these words lie in the nucleus of my cells that work for comprehension, they replicate again and again producing both types of effects on me; instant and delayed. Later on, when my little comprehension crept towards the greatness of Albert Einstein, Charles Darwin and Swami Vivekananda, and at least hundred others, I discovered how imagination was related with reality and how important was it to know the realities.

I began to live in the colony adjoining to the University campus. The cultures of cities were dissolving in the class. Open and secret analyses of attitudes were being done. It would fill me with joyous delicacy to have somebody asking me, "Sir, would you please tell me where such-such department is?" how interesting is it to perform a senior's task while being a junior! On few occasions, I would go with the askers to search the departments. We newcomers began to have our hair stand on end all day long due to ragging problem. New necks would turn all-round, junior eyes wide-open but unable to defeat the senior eyes. Finally, with mixed effects on hearts and minds, the first month of our post graduate studies passed.

University campus was very interesting from geographic point of view. Departments were constructed beside a hill. On the summit of this hill was a temple. Two types of people could be found moving up the temples stairs; those with the sole purpose of reaching the temple and those having a desire to enjoy the heights. On these stairs, I came face to face with the greatness of a man's purpose. For three days, I had been watching this old man carrying water in two big water-pots from down to up in the temple. I wondered, "These stairs make even empty-handed young devotees to gasp, how this old man is able to carry these heavy pots full of water!"

For once, I thought of offering him my help, but the next moment I identified myself as an utter fool. "O mother! My mother" his words were pulling me upward, I kept following him. Reaching the temple he transferred the water into a pitcher and began to serve the devotees. After observing him for three hours, I came to know at the time of evening prayer that he was the priest of the temple. Filled with joy, while I was bowing to him, he closed his eyes and sat down saying "O mother! My mother". The night commenced, I got down stairs.

Progenies
And progenitors
Are threaded
The journey is conceived
Life revolves around truths.

As yet, I was not close to any student in the class. There were two reasons behind this. One of them was I could not connect myself with the stylish livings of my classmates. I had not come from any big city like most of them had come from; Jhansi itself was a big city for me. The other reason was I had to study with English medium, and it propelled me to think day and night, "What would be the consequences of my failure? What about the pains that I would impose on the expectations of my parents

47

and Mr. Jain?" This year, we had to appear for three sessional exams and then the final exam, therefore, I got into searching the books that used simple language, and preparing the notes. I grew affection for two places in the campus; the library and the initial steps of the temple stairs.

One day, during lunch break, while I was sitting on those stairs, some of my classmates arrived. Among them, a boy with studious look whom, perhaps, I had noticed much less than the girls of my class, stood beside me and spoke, "Are you Deepak?"

"Yes, and you?"

"Rakesh Jain."

"Oho! Do you know Mr. Jain . . . I mean Mr. Rajeev jain? Where are you from?"

"No, I don't. I am from Lalitpur, it is not much far from here," he replied.

"Ok, come, sit," I said.

"Where are you from?" He asked while sitting.

"I have come from Banda, my family is there for few years, I have done my graduation from there itself," I replied.

He began to rub his spectacles with his handkerchief, peeped on this clean-up through the lenses and asked, "Do you live with any relatives or on rent?"

"On rent, in the colony adjacent left to the campus, Shivaji Nagar," I replied.

"Ok, I live in the colony right-side to the campus, Veerangana Nagar. Now I live alone, earlier, one of our classmates shared the room with me, now he has shifted away. My room is excellent, such rooms are not easily available here, if you like to join me, you may come to see the room." after this verbal use of several breaths and gaps, he began to look at my face.

"Why has your roommate shifted away?" I inquired.

"Thinking differed, he has been living in hostel since his secondary education, I like to live in family atmosphere," he replied.

"Let us go," I said. With few blows on the pants, we headed towards the room.

All along the way, we groped for each other's thinking on various issues. My nod became just a matter of formality before our entry in the room, this interactive walk had increased the dimensions of Rakesh's expectations. Upon entry I asked, "When should I bring my luggage?"

"Whenever you like, a fresh month will start the day after tomorrow."

"Ok, tomorrow evening," I finalized.

"Ok, come, let me see your present room, I will get there tomorrow to help you carry the luggage," he voiced firmly.

"Ok then, let us go, my room is two km away," I said.

Living together brought about a revolutionary effect on both of us. Rakesh's social understandings, his balanced, gentle and elderly ways prepared a home atmosphere for me. He stretched out the fragrance of my hidden hobbies and draped it all over me. It helped me to understand the significance of talent's diversity. Not even a single bud of my thinking could skip his eyes. In my absence, he would often say to our colleagues "Deepak is extraordinary, one day you will see him proving it." Often, I am afraid of such incessant praises of day and night as they help grow a big, hollow, poisonous fruit-bearing tree of arrogance spreading from ears to brain.

Rakesh was a diligent student. Every evening, after returning from University he would take a 30 minutes rest, then would prepare the tea. Leaving my share in the cattle, he would sit with his cup on his study table for hours. He had been studying on this table for years, to study elsewhere would give him a feeling of incompleteness. On my wake-up, he would utter "tea is in the cattle" breaking the sentence being read, and in turn, I would ask "Is the tea in the cattle itself in this chapter too?"

We differed in the methods of study. I would take a 10 minutes break every hour. In this gap I would imagine something with the things studied and then try to correlate my observations, keeping away from all the worries and pressures. Often, I would discuss different possibilities to Rakesh. He knew the importance of discussions, hence would keenly put his questions and give explanations, yet, finally these things would divert his attention, as a result of which many of his daily tasks would remain unaccomplished, making his target more difficult, puzzled, he would fall in deep thinking.

First sessional examinations were about to start next week. Our conditions became serious. We realized that our ways of study were not up to the need of rhythm. My problem was that I was dwelling upon the practicality of discussion, Rakesh awakened my past with Om wherein I had become the enthusiastic traveller of the path known as 'knowledge'. Rakesh's problem was that he could not underestimate the importance of a scientific discussion while later on he would conclude that he had made some mistake from the beginning. Studying for the first time with me, Rakesh began to quietly lose his confidence. It turned out to be a question of adjustment for us, a question of saving two right students from going wrong. Considering the flow of time, I decided to live separately for exam-time, Rakesh welcomed the decision.

Relationship is like a path which, on one side connects with the threshold and on the other side opens to the infinite, on it, persons and things do not appear or disappear groundlessly, they come, stay and leave according to the principles. One month later when I rejoined the room, we found that the past puzzles were rich in meanings, and we concluded that it is better to have a little time and small distance for a thinking rather than rashly implementing an opinion influenced by failures in maintaining the relationship, by doing so, the persons who have little or no interest in patience, may prevent the loss of good companions. The result of the first sessional examinations was out; Rakesh was in top three, I obtained the highest marks.

On September 11, 2001, Rakesh and I were discussing about the great contributors in the history of Microbiology when the plight of thousands of people entrapped between fire and height at World Trade Centre in America compelled us to wonder at the mean motives of a man who wishes to live a long murderous life. Man-made disasters! How filthy, malodorous and hollow at several places have they made the humanity! How openly the guilty Intentions have ridiculed at the innocent pursuits! And how frustrating is it to know that the people who are ready to die on the nods of their hidden bosses, follow them even if they are ordered to kill their own families including their own suicide! What an absurdity! Suicide bomber! Wake-up!

Brother
Remember
The days
When you held mother
Bursting into laughter
While eating you would say

"Food makes blood"
And when injured
"Blood brings tears"
You had a vision
You remembered the alphabets
Forgot the lessons
And songs too
Humming the cries
For a low price
You auctioned 'the self'
Poor bargain
Are you happy?
Braveheart!
For a fearless affection
Surmount the constraints
Shatter the terror
For a peaceful wall
For a happy door
Come home
Come soon

I wish man had not played with the simplicity of life. The complexity of life did not begin with the man, in fact, the life began before man. Was man at the beginning like he is today? Each of us now knows that he is an 'individual' with personal needs. We have intelligently reached in the present era of 'cultural legacy'. This cultural legacy has brought us to establish a simple relation between doing and being; do and be; what you do pushes you towards what you become. So far as the big question of life's relation with death is, there are several reasons to suppose that life and death are the prologue of the truth of which every living creature is an inseparable part.

Our studies were having a significant amount of disturbance these days; for several reasons. Our house was positioned at the centre of these reasons. A restaurant began to run nearby. Most of our acquaintances coming to the restaurant could not avoid visiting this house. Here, for past few days, two voices, completely disagreeing with each other, were coming from the adjoining house. Father's worrisome, whizzing, and whirling voice "No! You will not go on the roof every evening; don't you see the assemblage of boys on the neighbouring roof? Think of the consequences!" Daughter's sharp, whinny and vaulting voice "I will! Am I a child? If there are problems, there are solutions too". The problem was that the boys of our house differed in opinions and with the help of daughter's words, invariably reached on the roof to get involved in various activities.

We found it necessary to change the house. Entire experience was put in the selection of new house. This house was on the border of a village two km away from the previous house. This place was at peace taking guests. Also, there were two shops here; at one of these were available the dairy products and stationary related items, and the other was a combination of general store and electrical store. The people of the colony had intimate relationships with dairy animals and with one another. Embracing the diversity in this colony, both, village and city were populated.

In few days, we found out a shortcut to the University. The shortcut started with a five feet boundary hundred meters away from the house. This height of a barrier has the potential to attract a pathfinder. Much depended on the jumps. Soon, the classmates began to notice the frequent rips of our clothes, but whatever situation we were through, we never revealed anything about this shortcut.

Mostly, after returning from the University, we would spend the remaining time at the house itself. Often dreaming, we would construct the images of our future success through each other's eyes. I would say, "It is a dream, waiting to be proven as reality." Rakesh would say, "It is tomorrow's reality which you call a dream today." Every moment of our joint efforts turned out to be a torch bearer for our separate efforts.

The national game of India is hockey. Here, in Jhansi lived Major Dhyan Chand, the greatest Indian field hockey player. It is noteworthy that the glamour of cricket in India may confuse a person on the question of national game if he is not aware of the fact. The actual excitement of cricket in India is lived through discussions; in parliament, in offices, in laboratories, in fields, in jungles, in streets, and in houses. For some 15 days, on the University playground were appearing the students of Microbiology, Biochemistry and Biotechnology departments. Rakesh and I had been playing cricket since our childhood, old interests awoke, I hurt my knee, the play was over, now the thing left was a month's bed rest.

Three days later, Rakesh returned worried from the University. Recognizing an increase in the weight of his trouble due to my restlessness, he said, "Second sessional examinations will begin next week." and before my dilemma could be worded he said, "I know if the examinations were today, you would still be the top scorer, but the problem is of your uneasiness. The arrangement of vehicle has been made, few students have offered bikes, the classes will run for next five days, few girls were asking whether you need any notes prepared, I have given a nod to one of them."

Taking any type of preparation into account, girls differ from boys widely. They are concerned, cautious thus consistent by nature. Just think of two test conditions, in one of which the girls and the boys are given two subjects to prepare for a test without disclosing which one is the test subject. One may expect the girls, on an average, will secure more marks as compared to the boys. In the other test condition, both the groups are given to learn a short note full of alien words; one may note that the girls learnt the note more appropriately than the boys. Two points could be derived; first, girls don't take the options lightly, therefore have better backup plans. Second, girls, in general, don't tamper much with the original form of the things. These two points could be concluded thus: Girls are preparative; they better exemplify the functional totality; and it is reasonable to suppose that a girl was first in human history to understand that life was full of examinations.

Examinations were over and in a week's period, the result was also declared. We did better. I had feared for Rakesh's performance as he paid much attention to my fitness, but Rakesh balanced the things well. He was watchful to the happenings of my life. Happenings themselves were running towards me. I turned 20 and was learning to shave under Rakesh's guidance. The juvenile joy of being included in the shaving club rightfully with a full beard, and lack of hand perfection presented on my face a picture of their effects.

I, in hesitation, tried to conceal the fact from a girl that I had never driven a motorbike. The difficulty arose when she gave me the keys of her bike and said, "Can you bring my bike to a nearby shop, I will meet you there." Puzzled but committed, I rushed towards the bike at the speed of 'OK'. With the help of a person standing nearby, I unlocked the bike and dragging it for some 30 minutes, reached the shop. Frowning upon the strong sunlight seeing the girl's red face, I, without any delay began to search the part to kick in order to start the bike. The girl held the bike before I lost balance, started it and said, "Sit backside". It was since that day that I began to boldly say "No, I have never driven a bike."

The University was reverberating with poetic expressions these days; several poets had recently visited. Everywhere in the world; wherever you find the seeds of cultures, you will find the signs of great thinkers. The prose and poetic ways of expressions have played a key role in the development of human brain and thus civilizations. In India, both the ways of expression—the verbal and the implied are revered. Go through the history of world civilization, you will find how much respect is here for each and every language of the world. Does the name 'India' among several other names of this country like 'Hindustan', 'Hind', 'Bharatvarsha' and 'Bharat' not explain this? Each and every person of India is aware of the fact that languages are mothers and sisters. Think of a person of little knowledge like me who knows much less about English and whose mother language Hindi is guiding him to communicate in English, would one still say that Hindi and English are not sisters? The people who take pleasure in political tricks that aim nothing other than a mileage from creating disturbances, will, one day see their folly being uncovered. I salute to the poets and prose writers who have reached door to door to preserve the excellence of thoughts.

It had been two months since our third sessional examinations, and we were going to have the final examinations in upcoming few days. I had a sensational news from my parents; Mr. Jain had come to meet me at my Banda house. I left for Banda without any delay, kept busy in calculations all along the way: additions-subtractions on my progress and multiplications-divisions on Mr. Jain's progress. The next day Mr. Jain filled me with joy, he said, "Deepak, I have qualified the NET, soon I will join Central Drug Research Institute (CDRI), Lucknow for Ph.D. Anamika's father has invited me, we will soon tie a knot." Wow! What a journey! What else would better explain all that Mr. Jain and Anamika were through this year?

Mr. Jain instructed me to leave for Jhansi next day, "Don't stay here anymore, go and prove that you are the brightest in the University". Reaching Jhansi I got focused on studies like never before, but the circumstances themselves began to test us during the exams. With a day left to begin, all five papers were to be over within a fortnight. Beneath my lower lip was growing a pimple, it, within three days, turned my Asian lip into a Caribbean lip, and to open the mouth was very painful. It was a sweltering summer, the electricity supply got limited to eight hours a day, due to the failure of other measures, we were compelled to study in the heat of Kerosene-lamp light at night using wastepapers as fans. Our need of light made us to accept the presidency of insects, India is rich from mosquito point of view, Rakesh was allergic to the smoke of mosquito repellent coil, our tiffin service provider had a change of occupation, and we had to walk to a shop three km away, for all this, we were spending two hours twice a day, I caught fever.

Spending the precious time, medicines were arranged. Rakesh was balancing the things with too much softness, and this compelled me to say him, "Concentrate on your studies, if you lag behind, you will suffer more than we both are suffering today, I am fine, look at your aim." with the eyes full of tears he lifted my head up the bed and embraced saying, "I am very lucky", I pointed out towards his study table and said before closing my eyes, "Only studies. Bundle the emotions well and put them somewhere such that they don't fall before the eyes."

Rakesh now got into studies. A day and few hours were left for the exams. Difficulties were all around and on top of these was lying my diseased will. I prayed, "O God! Show me the way, I am unable to rise, take my hand in your hand, help me to step ahead." and I fell asleep. I had a feeling that I had heard something in my dream. I tried to recollect this 'something' many times but could never retrieve it precisely, but nonetheless I often assume it thus;

O, prince of trance
Weary and truth-touched
Treasure the remainder with the quotient
You are a student
Inaugurate the wonder moments
With evident aim sensations
Cogent, vibrant, constant
You are a student
Nurture the word seeds
Collect the meanings
Emboss the talent
You are a student
Pride of wisdom
Sculptor of definitions
The resident
Student

The incessant impulses awakened my will. The torrents of sensations picked me up from the puzzling shores and took to the canoe of efforts. A ray had reached to me that did not increase the temperature of my body, and its light did not attract the insects, the pains of body could not dare to appear before it, the time appeared to be in its rhythm. I learned the principle of preparation in these days only; the better you prepare, the more you enjoy the examination, next to this is success.

October 2002, marks and percentages were blowing everywhere in the University. Rakesh was at his home, he said on phone, "I will probably reach at our house by 2.00 pm, if I don't, I will meet you at the University, congratulations in advance for being the topper!" "How many times have I told you not to mingle your rashness with my worries!" I said. As I entered the department to obtain my mark sheet, an expectation-born fear took over me. Upon asking for mark sheet, my head of the department replied, "Take it from me after 15 minutes." When the presumptions fly in the wind of wait, the brain realizes the truth of lands again and again. The examination which made 30 minutes feel like three minutes, the 15 minutes of its result's wait made those 15 busy days of preparation appear so small.

After this long wait of 15 minutes, as I appeared before the head of the department once again, she extended her hand towards the bundle of mark sheets, smiled at my situation and said, "Let me see how many marks you have obtained, ok, 78.4 percent, congratulations! You have topped." Before I could understand what was actually going on in my mind at this particular moment, my Microbiology teacher Dr. Vijendra Mishra came and took me to the stairs of the department. He said, "Your style of accounting the things is amazing, elegant, you will receive a merit scholarship award of Rs 15,000.00". In this vocal music of appreciation and award, I asked him, "Sir, how many marks has Rakesh obtained?" "He is the other boy in top five, rest others are girls, Lo! He has come," he replied directing at Rakesh who was rushing towards us and returned to his room.

"I knew I was right. Haven't you topped?" Rakesh shouted from some 20 feet away. "Go, take your mark sheet first," I shouted in reply. As he returned I asked, "How are you feeling?" "I am feeling fine and relieved, what I had expected from you and from myself has fulfilled," he replied. "The mind boggling circumstances of examination time are now looking interesting, let us sit on the temple stairs, thereafter we will go to PCO and then anywhere else, I have to inform Mr. Jain and my parents," I said.

Success changes the forms of struggle; great changes follow to those who hanker after it. Twists, turns and supercoils are the fundamentals of life; they are not mere words that flash into the syllabus of a biologist, nor are they mere structures of a lifeless thread; they are the ways that define relationships. Effort and success are like two complementary threads that require only their relevant halves; this is the reason why a true student is with marks, a true artist is with admiration, a true politician is with the supporters and a true soldier is with his life, as much unconcerned as concerned.

4

Waves

The final year had a keen impact on the class. Priorities were being exposed. We had completely vacated the previous class for upcoming students and our seniors too, moreover, few students from both the pre-existing batches had passed the supplementary exams with a smashing total. Classes commenced from October 2002, final exams were to be conducted by the fourth month from the start of session followed by a few months' research project work called as 'Dissertation' at any of the research institutes in India.

Between the emotional farewell of seniors and the exciting welcome of newcomers, each and every student of the class was feeling a well known attachment. The attachment that never likes the separation of the past from the present and the future, the attachment for which the walls are as dear and lively as the life itself. The land where you grow and learn to explore makes you feel that strong gravitation which is by its scientific explanation always attractive, and this force, this pull, holds you tightly when you try to look at your fondness. This was the last year of our college life. Each of us, moving on the rim of college culture, was deeply concerned with the individual academic achievements, and friends. Few boys and girls had grown active in carefully distinguishing the 'friends' from 'more than friends' however, few of those once been considered as 'more than friends' were now in desperate need of at least a friend. Noticeable efforts were being made on relationships. Few boys were wandering here and there for the previous year class notes right from the day they offered to the newcomer girls to provide with them. They were found realizing how important it was to prepare the notes which they didn't! Few boys secretly forwarded the notes to the junior girls which they had obtained from their girlfriends, and were now in bidirectional loss since caught while few others were upset seeing their multipurpose help being substituted by some new help.

These days, Rakesh and I were having lots of bed-time chats. Mostly Rakesh would listen to me and I would confirm intermittently whether he is listening or sleeping, but on few occasions I would get engrossed in speaking unaware of this confirming and the next morning I would know that I had unknowingly woke him up by saying "goodnight", however, sometimes he would reciprocate saying goodnight in an intense voice and then fall silent as if someone had run for long just to bid a goodnight in my ear, then returned running.

How close the pieces of time sometimes appear! How simply the thought-waves bring the past, present and future before one-another! I, at bed-time, began to share with Rakesh the happenings of my early education, "As far as I can remember, my paternal aunt was my first teacher. She became widow in her early 20's. She had no child and would keep me with her all the times. She passed away some two years ago in her 40th. Due to her great care, preschool teachings, I got a direct admission in class 3rd; I didn't begin with class first or second. Often in school, my teachers and classmates would ask, "Who has taught you these lessons?" and I would reply "My paternal aunt, my mummy, my grandfather and my grandmother tell me these stories."

"Really, these tiny things of childhood make a lifelong impact," Rakesh said.

"Certainly. The lessons which I learned as a part of my entertainment, I see the modern childhood learning them in the pressure of performance in exams," I said.

"But the present day busy lifestyle is quite different from that time's lifestyle," Rakesh said.

"Of course lifestyles are different, the differences are even more in our urban population, but this is not a concrete reason, storytelling is less a matter of pastime, more of interest, parents are losing interest, I very well remember that my mother in our joint family, was never free from daily chores till late night, but she would synchronize her works with storytelling realizing children's curiosity," I said.

"Nowadays, parents have scarcity of stories," Rakesh said.

"I think it is better to say that parents have scarcity of stories that they can tell to their children, and I believe that if parents succeed in developing an interest for it, their sleeps will get better, their children will learn better," I said.

Bundelkhand University was established in 1975. During my education here revolutionary changes were made. Many departments were established. It is always a joy to see the growth of your institution with your individual growth. The ancient world had witnessed a University *Nalanda* in India being destroyed in an attack by a ruler shortly after the establishment of one of the oldest University *Oxford* and before the establishment of *Cambridge*. The Earth has seen numerous attacks on knowledge, the martyrs have always taught a lesson to the devastators that sacrifice for the sake of values is the stature of all human knowledge, and that humanity has nothing more straightforward to contribute in its own development.

The power-cuts and interrupted water supply compelled us to search for another house in the other colonies. We found no room having space for two persons; however, we found two adjoining rooms in a house. The rooms were on-road; one could have a direct access to the passers. Landlord informed us that a big room was under construction in the inner portion of the house which would take a maximum one month's time to complete. Since the house had no power or water restrictions, we shifted here.

After two weeks, one day in the evening, Rakesh shouted at my door, "I am going to a PCO to make a call at home, my room is not locked." I shouted in reply, "OK, bring sugar half kg." Some half an hour later Rakesh returned with a tensed face.

"What happened? Something wrong?" I asked.

"No no, everything is fine at home," he replied

"Then?" I asked

"Yaar, an old friend of mine has taken admission in the University; he wants to live with me," he replied.

"What is bothering in that? You tell him to wait for 15 days until the new room is completed, then he can shift with you there," I replied.

"And you?" he asked

"What 'you'? Don't think nonsensical" I roared.

"Ok, ok, I will tell him to come with his luggage after 15 days," Rakesh said.

Next day, Rakesh said, "Today I came across our classmate Dharmendra going somewhere with his luggage."

"Then?" I asked

"He said that his landlord troubles him so he has vacated his room and will stay with somebody until he finds a room. I have told him that he may ask us for any help in this matter," Rakesh said.

"All right, tomorrow we will tell him that you are leaving your present room in next 15 days," I promptly added.

By chance, few minutes later, Dharmendra passed through our house, we informed him. We came to know that he urgently needed a room, so much so that it would be very troublesome to him to wait even next 15 days, I lodged him in my room. Here, staying with me, often Dharmendra would wonder and sometimes go very emotional seeing affection of Rakesh and me towards each other. By his reflections on our friendship, I became sure of the fact that Rakesh's as well as my brain works better in each other's matters.

Dharmendra tried to conceal at his best that he had no money but I smelled that he would not be able to arrange money even in the next three or four months. I thought might be there was some very serious reason for he was unable to speak of his predicament. After a day of thinking I came up with an Idea. I discussed the idea with rakesh, "See Dharmendra is concealing his financial problems, he is hesitating to ask our help, maybe he is thinking that we have already rendered enough help to him, now the solution is—I got Rs 15,000 as scholarship unexpectedly, I still have Rs 5000.00 which I may indirectly extend to Dharmendra."

"The idea deserves you. What plans do you have to let it all happen?" Rakesh asked

"I will plan the whole thing soon," I replied

Next day, I discussed the idea with my teacher Dr. Vijendra Mishra. He said, "Ok, meet me tonight at my house; we will develop the idea and will extend the amount in the form of some special scholarship etc." "Ok sir, should I come with the money?" I asked. "As you wish," he replied.

After discussing the whole thing with Rakesh, I took money and left for Dr. Mishra's house. Some 45 minutes later as I was approaching his door, I found my pocket empty, I had lost the money somewhere in the way, I had first time lost the money in the way, informing Dr. Mishra about this loss I got back to my room, the incident left Rakesh dejected.

How much lies between an idea of help and to succeed in helping! Something that sometimes appears to be as small as a four letter word, and sometimes as big as its meaning. Help simplifies life. When an individual learns to harmonize with his internal and external milieus, he comes to know the essence of help and then he achieves the satisfaction he has been looking for. How painful a failure to help is! And how intense it grows when you, while walking for some urgent work, happen to see an animal helping an animal.

I determined that I would not let Dharmendra go away. Rakesh shifted with his friend in the newly prepared room. My room was too small to adjust two persons comfortably.

I found a room, some 100 meters away from the present room and shifted there with Dharmendra. Here, almost twice a week, one of our classmates and Dharmendra's friend Ramgopal visited to meet with him. Ramgopal was a very busy boy, for two reasons: first, he was a very affectionate person and second; he lived in a big house crowded with boys of different disciplines.

One day, Dharmendra returned from the University with a serious look. It was always very difficult to make Dharmendra speak of his problems. Having been insisted for hours, he confided at night that one of our girl classmates had addressed him as 'brother'!

"Nowadays . . . Girls don't use this word outside their families unless they get into a situation that they can no longer avoid to do so; the other reason may be 'a slip of tongue'," I reasoned.

"No, no. No slip of tongue, it is deliberate," Dharmendra fumed.

"Ok! Then surely she needed your help. Sometimes, in clean heartedness, person tends to forget such calculations," I promptly added.

"Help! Who is running away from help? I will surely help her but you know very well the ways of girls, now you see how many girls call me 'brother', and then boys will make fun of me" while saying all this Dharmendra's cheeks swelled up like bread buns.

"What fun will they make? Can anybody dare to call you 'brother-in law'?" I wore a serious outlook.

Dharmendra got relaxed, he saw here and there in the room, and then began to carefully read a book.

Ramgopal was in search of a study atmosphere since he took admission in M.Sc. last year. He wanted to spend time studying seriously. He began to linger at our room.

His questions once again awakened my passion for a systematic discussion. For a systematic discussion on any subject, it is very important to understand the available facts and associated principles. Ramgopal now wanted to keep studying for long, he began to leave for his room late at night, just to pass the night there, and he would join us early in the morning. His displeased housemates would thoroughly taunt at him. Ramgopal was never affected by these things because he well understood the hidden affection of the housemates, and the boys even while expressing their anger, would always remember his likes and dislikes.

India is rich in family values. Here the roots of relationships are protected. Protected in two ways: ensuring that there is at least proper air and water, if not sufficient mineral nutrients, and ensuring that growing roots are not going to crack their foundations. Friendship has explicit family importance here but like anywhere else in the world, a friend may be found in a famous dilemma of friendship which is—There are three friends A, B and C, very dear to one another but while A is with B, C becomes alone, later he begins to feel something lacking in his individual friendship with A and B. I have found many close friends getting into a helpless situation taking this real problem. Is a square better than a triangle in relationships? Is somebody D needed? I have two observations: An Urgent move to have somebody D to feel better is not effective in a long run, and those who have an unbiased silence along with faith in their friendship find an appropriate D without losing A and B.

February 2003. The final examinations were over, now we were looking for the research institutes to have our few months' dissertation to be conducted at. Many students with their forwardal letters left for big institutes, few of them even started their work. Rakesh and Ramgopal left together for Delhi. I also, for the first time, prepared my bio-data, but where would I go? I didn't know. One day Dr. Mishra called me and said, "I have a dissertation offer for two top students of the class. The offer is from All India Institute of Medical Sciences (AIIMS), New Delhi. I will also talk to the other student; I hope you will soon get there."

I was the happiest student of the class at this moment. "Sir, your words are best-timed, I desperately needed these words, thank you so much." I sang in glee and began to search a point where I could sit alone to imagine myself at AIIMS. What more than AIIMS could I even imagine? Few hours ago I did not know of any place where I could at least send an application for a dissertation, and now it was New Delhi to be put as the destination in the railway reservation form! My imagination finally landed at Jhansi railway station.

Rakesh and Ramgopal were back with good news from New Delhi; Ramgopal got selected at Jawaharlal Nehru University (JNU) and Rakesh at Centre for Biochemical Technology (CBT). CBT is now known as Institute of Genomics and Integrative Biology (IGIB). The institutes whose names we had been hearing from our teachers and seniors for long, we were going to be the part of their upcoming history. I began to imagine various kinds of institutes, research laboratories and researchers absorbed in their research, but among these images were the fields with no buildings or boundaries. These imaginary fields were waiting to construct the structure of an institute where Dharmendra would carry out his dissertation; neither he nor I knew the institute yet. One day, some 5 days later, these plains suddenly disappeared inside the huge buildings when Dharmendra came gasping and said, "Deepak, I am selected at Forest Research Institute (FRI), Dehradun."

I was thinking of Delhi, "Delhi is so busy! My friends say that people rush to catch the buses; buses have stoppage time of few seconds, I have to search for everything there, AIIMS, room to stay, research methods, helps, I will surely get helps, Delhi people are helpful, but I will have to do everything by my own, I have enough to learn, I need to be bold."

"The stay arrangement has been made, there is no problem at all, I have already paid one month's advance for a room there," Ramgopal's voice surpassed the noise of my thoughts.

"Is it! Where to stay?" I asked following the tilt of my neck.

"Near to AIIMS and JNU is a colony 'Ber Sarai', the hub of students, I will take you there, we will leave next week, book the rail tickets," he replied.

We arrived with bag and baggage at New Delhi railway station. It was raining heavily. Ramgopal said, "You stay here with luggage, I am going to find out which bus goes to AIIMS." "We have to go by bus no. 56." He said on returning. He handed me two light bags and got loaded with the rest of the luggage, moved ahead of me to the bus. At the main gate of AIIMS he said, "Let us walk straight, to the building before us, we will ask somebody there about the department you have to reach, then I will stay there with the luggage until your meeting with your mentor is over." Completely agreed with his point, I walked with him towards the building, I said, "AIIMS seems to be bigger and busier than that I have seen on

TV, Have a look upward! How short a life and one's own stature appear moving before the 'Emergency'!" "Yes dear, the luggage appears to be weightless seeing the old family members taking care of their diseased young ones," Ramgopal said.

"I am not interested in the marks you obtained or scholarships etc you received because you know your University is not among the leading Universities in India, I have given you this opportunity because your Vice Chancellor is my friend. I don't want to see you working here with that level of attitude, now you are at AIIMS," said the AIIMS mentor.

"Write an application, get it forwarded from me and submit the fee today, come to me tomorrow," he added.

"OK Sir, Thank you very much," I replied and rushed for fee submission.

"You have been busy with your mentor for quite a long!" Ramgopal worded out his long running thought.

"No, the meeting was very short, some 5 minutes or so, finding out the accounts office and the submission of fee took much time," I replied.

"OK, how was the meeting?" Ramgopal asked.

"Perhaps motivating, I came across new high expectations," I replied.

"And I am sure you will meet them well," Ramgopal's words relieved my nervousness.

"Our room is not much far from here, since we have luggage, we should take an auto, we will have it at the main gate, you take care of your documents file, I will carry the rest," he added.

"Why you? I can carry two bags," I fumed.

"Not needed dear, put it all on my shoulders, you don't know I have learnt this use of shoulders from my house servant who used to carry me to school on his shoulders, in my childhood," he said stepping fast.

JNU is at a walking distance from Ber Sarai, for AIIMS you need to have a vehicle. On our day first as students of AIIMS and JNU, while leaving for our respective institutes, Ramgopal said, "Bus no. 507 goes to AIIMS every half an hour. We have to go opposite to each other, first you catch the bus, then I will leave for JNU, buses take a very short stoppage, take it like they don't stop, just slow down and passengers rush haphazardly up and down, the first door to get in and the second one to get off, take care." "Get back by the same bus in the evening, it passes through the AIIMS gate," Ramgopal shouted as I got into the bus, I gave a wave of my hand indicating "Ok, ok, I got it, don't shout, relax." he waved his hand in reply meaning "Bye dear."

My first day at AIIMS was not as easy as I had imagined it to be, though the beginning was excellent, the mentor introduced me to the lab. The lab had a senior resident, two MD students, five Ph.D. Students, four research assistants, five regular technicians, two data entry operator, two attendants and four M. Sc. Dissertation trainee including me. Here, genomic, proteomic and immunological studies on microorganisms were being conducted. I was not given any topic as of now but was told to collect the literature first. In a few hours, I began to realize that I was not being welcomed; I found that I was being perceived as an unwanted element to a well set atmosphere of the laboratory.

In the evening while leaving the Institute I saw a research assistant of my lab moving ahead of me, I approached him and requested, "Sir . . . Sir would you please tell me where bus no. 507 passes through, I have to go Ber Sarai" "Hmmm . . . . Ber Sarai! . . . . I will show you the spot" He said without turning his head to me. "Thank you sir," I said and increased my pace. Suddenly he began to run along a bus, I followed him but this was not 507, I saw him catching that bus, I stopped and began to look for my bus. The dusk commenced. After some 15 minutes or so 507 arrived and I jumped into it. The helping jolts of passengers positioned me somewhere at the centre of the bus. Some 5 minutes later, conductor approached me hustling the passengers. I said, "I have to go Ber Sarai." "Then what are you doing in this bus? Take the bus from the opposite side, get off at next stoppage, give me 5 rupees," he shouted.

Finally, I got the right bus. It struck 9:00 getting Ber Sarai. But I got utterly confused about my position; I was not getting any clue for my street from the corner I was walking in, and I could not seek anybody's help due to a feeling of embarrassment for I did not know the address of my room. I decided to keep moving into the road I was wandering on. After a 100 meters walk I saw Ramgopal coming. "Where were you? Do you know I was searching for you since 5:30 on this road, I saw a boy just like you, wearing the clothes same as you in a bus, I thought perhaps you have forgotten the stoppage, so began to run behind the bus, fell down," said Ramgopal. "Oh! You are injured, clothes are torn, palms and knees are bleeding, uf!" filled with tears, unable to speak further, I became silent. "It's nothing yaar. First you tell me why got so late? Heavy workload?" He diverted my mind. "No no, I left the institute at 5:30 but made some mistakes in the way, leave it at all but see you are injured!" I said. "Stop thinking about it, this is temporary, let's go for a tea, I was waiting for you," he said. "Why don't you activate your mobile phone? You are the only student in the class who possesses it since the beginning of the course," I said. "It will take some time, I need to purchase a new SIM," he replied.

Though I had gained enough confidence, Ramgopal, for the next 5 days would be with me until I catch the bus, and then he would go to his institute. Within three days in the lab, I came across my backwardness. Seniors' words were heaping up in my mind. "Listen! Whatever your name is, yes you; don't try to touch any of the laboratory registers and records." "These trainees are a menace; they come for their own benefits and create a hotchpotch." "Don't touch the glasswares or instruments without permission." "Your help is not need, keep away, there." "Where is Bundelkhand University?" "Don't try to sit in any of these chairs, arrange stools." "What is the formula to calculate molarity of a chemical solution?" "Why do you pass whole day here? Can't you go library or some other place?"

To pass a day full of such fresh remarks was very painful. The evening would see my dismal face and Ramgopal's rage "Why don't you tell your mentor about all this?" One day I decided to go to the mentor. As I appeared before him, he said, "You will have to purchase a separate micropipette for your work, you people mishandle the micropipettes, the lab suffers, convey this to those three girls too." (Micropipette is an instrument used to accurately measure and dispense small volumes of liquid; from less than 1 to 1000 microliters). "Sir I will carefully handle the laboratory micropipettes," I requested. "Then go back to Jhansi, there is absolutely no need of you here, you argue," he said. "OK sir, I purchase it, where can I get it from?" I asked in a low voice. "The supplier will contact you; he will get here next week," he softly replied.

I realized that I needed to be patient "I should take it easy, someday, somebody would understand me, after all, they are seniors and if they are satisfying their egos, then on an easy day they would extend their care, and this is a part of my learning, sometimes it is this way and sometimes in a more different way, but micropipette is costly and I receive money from home through money order, don't know how much time it will take to reach me lest I should go back with bag and baggage to Jhansi". Amid the heavy traffic of thoughts I began to wait for evening.

In the evening, upon entering the room, Ramgopal asked, "What happened? Have you informed your mentor about what is going on with you in the lab?"

"Yes I went to him, but his points suggested that I should not tell him all this," I replied.

"Is it? What did he say?"

"He told me to purchase a micropipette," I said.

"It is so strange that a trainee is told to purchase a micropipette for his work, didn't you request for if it was possible to work without buying it?" Ramgopal said.

"I made a request; he got annoyed and told me to go back to Jhansi," I said.

"Your path is very challenging dear," Ramgopal envisaged.

"Have you seen me having an easy road?" I confirmed.

"Take this money, a micropipette would cost around rupees 4000, I know that you have no money right now," Ramgopal said taking money out of his briefcase.

"These are 5000," I said.

"What if you need them too?" Ramgopal began to look at me.

"All right, I keep it all." I put the money in an envelope and placed the envelope in my bag.

I, along with the rest three trainees submitted the micropipettes to the lab. I had completed a month at AIIMS. While leaving Jhansi, I had no idea of whatever I went through in this past month. I could not mingle with the laboratory persons; in fact, I discovered a diffident in me. My growing dark fear abducted my infant amiability. Whenever I tried to have a scientific discussion with anyone, I was replied in a way that sounded like I was at an unpardonable fault. Yet I had no serious problems with anybody here because I realized that each of them was under certain pressure.

The AIIMS mentor had an excellent quality with two evident effects: He would every third day evaluate the progress of the lab and would allot the topics to at least two students a week to prepare for the presentations. He always followed the question-answer method of learning. A shower of questions would occur all around no matter who was the speaker. We trainees would be the best choice for an interesting, refreshing target, however during such inevitable rains we would put our best to hide behind our juniority; Adjusted between the last row and the wall.

I became fond of B B Dikshit library of AIIMS. Nothing else was better than getting here; this place not only welcomed me but also offered me the opportunity to learn peacefully without bringing my dignity into the question. In addition to learning from the books, I learned much from the readers with whom I never had a verbal communication. Though National Medical Library (NML) neighbours AIIMS, I went there only on two occasions. Here, at B B Dikshit library I found as if I was a tiny cloud passing through the shining invaluable gems, really, you look at a person lost in his work, the pervading innocent dedication will hypnotize you.

The library improved my performance in the laboratory. I was answering well, was gaining confidence. Confidence is exponential; its ups and downs are characterized by the dynamics of will. The second month was about to move past and the mentor told me that I would identify some proteins by western blot technique, moreover he directed to a Ph.D. student to guide me in all this. I wanted to collect all the required materials before the end of the month. Since I was told to develop all the operating protocols on my own, I also began to visit at computer facility.

Until the first weekend of my third month at AIIMS, I gathered all the required protocols and reagents, but the Ph.D. student whom the mentor had told to guide me, was writing his thesis, and used to get very tensed, frequently flared up. Realizing that neither he needs my help nor am I so able to help him, I kept on waiting to come to his attention. The month would be over after a week's time; the tenure of my dissertation would end next month.

I began to get very upset, seeing my situation, one of the laboratory research assistant, one day, consoled me, "Deepak, I know that you are in trouble, I can understand your problem, I have also been a trainee here, I will try to help you, but not directly, it may affect my personal relations with all the laboratory persons and with our mentor also, you know, nobody will take it positively, because I am senior to you but also a junior fellow to most of them, after all." "Thank you sir, your words are very soothing, I will remember these words," saying all this my voice began to tremble.

In adversity, even a dwarf riding a camel is bit by a dog. On an afternoon of the passing month, upon returning from the library, I saw a laboratory computer's screen moving haphazardly. Now, as I held the mouse the computer got off, I didn't try to restart it instead informed one of a senior laboratory person. The matter reached to the mentor very quickly. He called me up and said, "No one else has tampered with the computer, I don't need the students like you, leave my lab right now, go back to Jhansi." "No sir, I haven't, the screen was already moving haphazardly, I only held the mouse," I pleaded. "I don't want to listen anything, Give me in writing that I had last operated the computer before it stopped working, whose entry was not made in the logbook, and get lost!" he said. I followed his order and said, "Sir I am doing all this for a scientific spirit, please do look into the matter once, I will leave the lab in next 5 minutes."

With the eyes full of unjust reality and tears, I left for my room. Instead of going room, I wandered in a field near the post office, then sat on the ground under a tree. Fighting with the thoughts, I began to whimper . . ." How these people are! Why do they do so? What my fault is? What wrong have I done to anybody? God! Do you see? God . . . . O God! . . . God . . . . O God! . . . . see God see! What they are doing . . . God . . . . O God . . . ."

This happened, Oh! This happened!
True! Yes! True
And wrong, Ah!
Thus embraced my grief

The evening progressed fast, it was 9:00. An old man came near to me, gazed upon, and then went away. I took my bag and came to the room. Ramgopal was not back yet, I got worried, what to do? Began to look here and there in the room, found a note "Dear Deepak, enough is left to get finished, therefore I will stay in the lab tonight, I had prepared the dinner at 6 PM, have it before it spoils, I will get there early in the morning."

Ramgopal woke me up at 8 AM. "When did you come? How did you get in?" I asked in a surprise. "The door was open, I got here at 6 o' clock, I have even taken a bath, breakfast is ready," he said.

"OK," I said and began to think of yesterday's events.

"Hello! Where? Get ready, don't have to go lab?" he said.

"No. I am thrown out of the lab," I replied.

"What? Why?" he asked.

"Because the mentor believes that I tampered with the laboratory computer, while the thing happened was—upon returning from the library, I saw the computer's screen moving haphazardly. As I held the mouse the computer went off, I informed one of a senior laboratory person, within an hour the mentor took this action," I replied.

"How unfortunate! My lab is very peaceful, no one behaves like that. You see I know nothing but nobody has treated me like this," he said.

"I just get back from my lab in an hour, I should inform that I will not come to the lab for today," he said.

"Why? Why are you dropping today?" I asked half-heartedly.

"Yaar, I was in the lab whole night, should I take some rest or not? Otherwise I may fall ill," he lied neatly.

"You have learned well to beat around the bush," I said, he left with a smile leaving me smiling.

Now, I was left with a month's time and in this time I had to accomplish a very tedious task: To find a different lab, to have a fresh work, and then to finish it with an acceptable conclusion. The goal was merely impossible, but I wanted to avoid the worst. I said to Ramgopal, "You don't stop your work, you have enough to complete, let me first try at few places." Searching for a laboratory, I reached at CBT and met with Rakesh. Rakesh got very upset with my position, he suggested me to try in a lab adjoining to his lab. According to his advice, after a few hours wait, I knocked the door at 7 PM.

"May I come in Sir?" I asked.

"Yes," a man, aged some 40 replied.

"Sir I am Deepak Kumar Dwivedi, a M.Sc. final year student from Bundelkhand University, Jhansi."

"OK," he provided me a push to connect further

"Sir, I want your help to carry out my Dissertation, I seek your kind guidance for a month's time," I added.

"Dissertation in a month's time! Do you think it is feasible?" he asked.

"Yes Sir. Now I am left with this period only, I tell you the fact behind this. I was already working at AIIMS since February, I am turned out of lab," I replied.

"Why? Who was your mentor?" he inquired.

"Sir, to be honest, I don't want to tell his name to anyone, because I don't want anyone criticizing the mentor due to me, but yes I would not hesitate to tell that he turned me out without any inquiry, I am not a culprit, and if needed, I can narrate you the reason for which I was turned out," I replied.

"Ok, no no, no need to mention the reason, actually, I have no space, so . . . I am sorry I cannot help you." Then he began to turn the files put on his table.

"Thank you sir," I said and returned to my room.

The next day, Rakesh took me to a Ph.D. Student, he said, "Here, nearby are three institutes—Defence Institute of Physiology & Allied Sciences (DIPAS), Malaria

Research Centre (MRC) and Vallabhbhai Patel Chest Institute (VPCI). I don't know anyone at these institutes and I would suggest you to try at these places without any reference." I first tried at DIPAS but the problem remained as it was. At this weekend, I got terribly tensed. Ramgopal said, "You know astrology, try to find out the cause of these problems."

"I had no such Idea. I don't even have my birth details, and I have not practiced the astrological principles for long, My astrology Guru once said that after him, I would make use of this science for the good of people and I have grown oblivious of all this, I am a stupid," I said.

"Very few people naturally get such chances; do you think it is without a purpose?" he said.

"Yaar, I believe that everything is according to a grand principle. There are events that can be predicted, and there are events that can never be predicted. Merely countless things are part of a life and a life is part of merely countless things. Yet, in the sphere of possibilities, I have come to believe in 'chance', and it is this chance along with the fundamental universal principles which everything seems to be the product of, to predict this central event is the biggest challenge for any kind of knowledge, I think," I said.

"A friend told me of an astrologer who lives in Shahdara colony in Old Delhi, we should go to him, you may also learn something from him, call at your home and get your birth details."

"Ok. When should we go?" I asked.

"Tomorrow," he said.

Next day in the afternoon, we began to search the astrologer's whereabouts. People informed that he sat in a temple, we found the temple. In the courtyard, on the mats were sitting some 20-25 askers. In front of these askers was sitting the astrologer whom the askers were addressing as 'Pandit ji'. Questions and answers were going on. Some were looking in more trouble than me while the others were in less, few others were discussing about their future plans for a better development. We had passed an hour or so. It was 6 o'clock; it was the time for evening prayer. A disciple reminded Pandit ji for the prayer, we all attended the prayer, many of us were praying except Pandit ji and me, he was looking here and there and I, at him all this while.

Just after the prayer, I covered my mouth and Ramgopal's ear with my palm and whispered, "My devotion and my reasoning took different directions seeing Pandit ji, let's go back."

Ramgopal gestured in 'why'

"It's done, nothing more to know from him," I replied in a low voice.

"Hey, let's have a single advantage of this whole exercise," he said.

"He will give us nothing," I said.

"Just wait, two more persons to go," he insisted.

"Yes! What's the problem?" Pandit ji asked.

"I am a student, I was working in a laboratory. I had only one month left to complete my work. Due to some circumstances, I could not even start my work in past three months, now I am turned out of the lab. I am trying to find a different place, nobody is taking me, what should I do?" I said.

"What was the reason of the termination?" he inquired.

"He thinks that I tampered with the laboratory computer, which is not true," I replied.

"When did he do so?" he further asked.

"5 days ago," I replied.

"He has now prepared the papers against you," he said after some calculations.

"Is it! But why would he need to do so? I am a kid before him, I had not even resisted for anything," I said.

"Yours is a government system and in a government system nothing happens without papers," Pandit ji said with a smile. Pandit ji's insensitivity was exposed to me during prayer itself, I offered some money to him and indicated Ramgopal to move. As we stood up, different suggestions from the askers began to float "Go to the mentor and request for your redemption." "Touch his feet and say that your career is at stake." "Purchase a new keyboard and give it to him, keyboard doesn't cost much." "No no, it's a problem of RAM (random access memory), replace that with a new one." "Hurry up, Ramgopal!" I said and breathed a sigh of relief on exit.

"You were right, it gave us no advantage coming here," Ramgopal said.

"Such people do severe harms to astrology. A subject of scientific and social importance is thus decaying due to such mentalities. You see, the misuse of commonsense psychology was apparent in the astrologer's remarks. It is not tough to sense the magnitude of pain in a person having a hard time, he bursts-open on a slight touch of an inquiry, and people like this astrologer fabricate the rules with some tentative correlations, and then put all this in a very mysterious way," I said.

"You are right," Ramgopal said.

"These selfish people encash the fruits of explorations made by ancient virtuous seekers of truth. Look at TV channels! They are everywhere, with their real-looking fake versions on various mysterious lives," I said.

"Righteous people keep away from such things, they live a very simple life, and they are so by nature. Why would a knower of truth need such tumult? I have read the history of India, and I believe that the sacrifices and donations of the great people will never be worthless. Look at the national emblem of India, It narrates a great experience made perceptible as 'Truth alone Triumphs'. This is the declaration of *Upanisads*. Truth is the soul of India," I added and became silent.

"I am few years older than you but now it seems that I am a kid, why have you gone silent? Continue," Ramgopal said.

"What to take of the age? Let's find a bus," I said.

"Why don't you discuss such things daily?" Ramgopal said.

"It's great to know that you have interest in such matters, Rakesh is fond of all this, but I find that the present day's youth takes so little interest in the systematic study of social and religious matters, with his raw belief, he is lost somewhere in fashionable technologies and few distinct youth matters, he wants to delay in accepting that he carries the hopes of future, he should be able to distinguish between fake and true like we saw here, later he will grow older and fakes, mightier," I said.

"I feel something new, when you speak, I feel that I am speaking, even knowing that I am unable to think like you," Ramgopal said.

"Do you know you have just worded something very special? I will remember these words," I said.

After reaching the room, Ramgopal asked, "Do you find any of the temple asker's suggestions considerable?"

"No, however their suggestions were practical and far better than Pandit ji's remarks," I replied.

"How would it be to meet the mentor once?" Ramgopal slowly projected one of the devised options.

"To touch the feet and to feel utterly sorry for the mistake I couldn't make? The one who is aware of the self-respect always readily concedes that he has made a mistake, but he cannot pretend to regret. Do you know what this Indian custom of touching the feet means? It is the symbol of respect and devotion; it is not an exercise of flattery, it is very unfortunate to see this great custom being terribly exploited in India," My voice became louder.

"In any civilization, great customs never disappear due to disasters; they fall diseased, then buried alive by murderous hands," I added.

"Well said, I completely agree," Ramgopal began to shake his head.

"What is tomorrow's plan?" he asked suddenly holding the head.

"Mall road side institutes," I replied in a low voice.

Next day, I set out to Mall road but on reaching the nearby stoppage, I felt a lack of confidence in me, dropped the idea to go anywhere, came back to the room and fell asleep. In the evening, Ramgopal informed, "I had a call on my mobile from a boy of your lab; your mentor wants you join the lab tomorrow."

"What?" I said.

"It seems that he has now come to know that you are innocent," Ramgopal reasoned.

"May be, let me prepare for tomorrow's adventure," I said.

I had always walked through a subway connecting AIIMS to the opposite side stoppage. Now I was feeling bad to pass through this subway.

"Where were you? Do you know for how many days you have been absent?" The mentor asked.

"Sir, I was searching for a lab," I replied.

"Go and do your work . . . . 'was searching for a lab' researcher!" Repressing my pain, he ordered.

"OK sir," I replied and left his office.

Rejoining the lab, I began to think "What should I do now, and how? The arrangements that I had made to start the experimental work before my ouster are now traceless." Another problem cropped up, the Ph.D. student whom the mentor had told to guide me, was on leave for next 15 days. My project title was not given to me yet, and it was not looking possible to get it in near future. Who would do this for me? I prepared a title and objectives of the study, and went to the mentor.

"May I come in sir?" I asked.

"No. Come after 5 o'clock, what is the purpose?" he asked.

"Sir, I have prepared the title and objectives of my study, looking for your approval," I replied in nervousness.

"Leave it on my table, come after 5 o'clock," he said.

"OK sir," I left the papers on his table.

At 5 o'clock, I reached his office again, he was leaving the office. He said, "Take those papers from the table and finish your work at the earliest." "OK Sir," I replied. I was happy for my proposal was approved, but the experimental part was a worry. I had never seen anyone performing the experiments that I had to run. I was in this maze when a lab person said to me, "You don't know, I had requested to the mentor that Deepak is not a culprit in the computer case, this is why he has called you back, now the culprit will not be pardoned."

"Nobody is culprit, even the computer is not a culprit, is it necessary to take every mishap as somebody's fault?" I reacted.

"Do you want to stay in lab or not?" he posed a threat.

"I will not linger," I corrected his calculation, he saw here and there, then moved away.

The laboratory persons sympathized, none of them rendered any help. Ramgopal was relieved to some extent as I was permitted to rejoin the lab, but he was in the firm grip of my predicament, he said, "A Ph.D. Student of my lab is doing somewhat similar kind of work on proteins, She is from Nepal, I will request her to give you a demonstration, I will also take my mentor's permission; he will not prevent you from learning anything in the lab, he is a very gentle person."

Ramgopal's generosity, like always, clenched-shattered my trouble. An hour of affectionate demonstration by that Nepalese lady named Dr. Anjana, whom I never met again, showed me the way to conduct my experiments. Since the experimental setup at my laboratory differed from that of hers, I, for the first time in my life concluded that one of the fundamental steps in science experiments that are performed with instruments is 'standardization'. After few difficulties, I completed my work, though the lack of time was printed in my project report, nonetheless, these few months had mentored me thoroughly. While leaving, another good thing that happened was that I got everyone's support in the lab during this last week. In India, the students never leave their institutions without offerings; they are fond of a feeling of self-satisfaction that implies that their offerings have been accepted, the laboratory had accepted the micropipettes from each of us trainees.

Deepak Dwivedi

We completed M.Sc. in September 2003. I, with Ramgopal and Rakesh stayed in Delhi. The Ber Sarai room was too small to adjust three; we found a bigger room in a nearby colony Munirka. Fifteen days were to go to end this month. As we informed the landlord about all this, he flared up and said, "I will not let you go until paid an additional month's rent, you must have informed a month ago." We requested, "Sir, you never informed us about any such terms and conditions, moreover, we are leaving the room with 15 days left for which we have paid, for your satisfaction we will also pay an additional 15 days rent." "I have nothing to do with when you will leave, I will not let you go until you pay for next one month after this month," he thundered. We tried to persuade him, he said, "I will break your lock and throw your luggage on the road." We decided to take police help. I said Ramgopal and Rakesh, "I am going to find out if there is any police station nearby, you both stay here; he was talking of throwing the luggage."

I found a policeman at street corner. I requested him for help. I, for the first time in my life saw a policeman having his self-esteem clung to his chest while on duty, standing with his colleagues whose self-esteem and duties were peeping out of their separate pockets. Years after, today, while I think of that evening, I wonder, how fortunate I was that I came across that honest policeman about 50 year old, addressed as Sharma ji!

"Don't worry child, take me to the landlord, but I foretell you that I will persuade him saying that I have acquaintance with you, maybe he become somewhat soft. But a stronger possibility is that the situation will get complicated, I know the attitudes of these people, he may even try some warning signals to me giving reference of my seniors, and maybe he knows them, therefore during my enquiry, you get your friends ready for Hauz Khas police station. There you file an FIR (First Information Report). If I leave you here, any mishap may occur, and you may be victimized, these people are unpredictable," Sharma ji suggested. "OK Sir, thank you so much, really," I said and followed him.

The landlord was somewhere outside, his wife came out rushing towards us with her three year old son as if she had seen a policeman at her door from somewhere inside and launched an attack at Sharma ji, "How dare you stand at my door! Who street-policeman you are? I will file a molestation complaint."

"Either you speak in manner or if you cannot, leave your son in a peaceful room, do you know he will begin to cry in a moment? Listen one thing carefully, send your husband to me, I will be available at street corner for tonight, and don't forget to note that neither these children are helpless nor they are in a strange city," Sharma ji blunted her attack.

She kept jabbering, we along with Sharma ji, came out of street. Sharma ji said, "You go Hauz Khas police station and wait for me, I will reach there in half an hour". It was 9 PM, we reached the police station, wrote an application and waited for Sharma ji. He came, introduced us with one of his colleagues who received the application and returned to his duty area. His colleague said, "You see, many students like you are behind bars, most of them have same issue, they were unable to find any support, they are studious, few may be seen studying day and night, it's sad to see them here."

"Sir, we don't want to go our room tonight, landlord's wife was threatening Sharma ji of molestation, can we stay here for few hours?" I requested.

"OK," he said.

We planned to leave at 6 AM. Meanwhile, at 11 PM, an event of staff member's recreation took place; three well clad eunuchs (generally called Hijra and Kinnar in Hindi) appeared at police station. Both the sides were engrossed in talks of mutual pleasures till 4 AM, and we, pretending to be unconcerned with the matter, remained embarrassingly concerned with the essence of the event.

Is our society awaiting an ideal eunuch? Has the world ever honoured a eunuch? Have you ever been touched by a eunuch's greatness? If there is "Yes" anywhere among these possibilities, the society must know the meaning of 'a true eunuch'. We must ensure that the definition of a true eunuch is not solely derived from the inferences of sex-oriented physiology, but also from the physiological reflections that can express the heat content of great morals. A true eunuch gives blessings to the society for the attributes which he lacks. He never takes the path of vulgarity, cursing, assaults and prostitution, but he treads a path of fertile equality. A society can never deflect a eunuch; he owns a masculine endurance, feminine patience.

A eunuch like anyone else has a mother and a father. Grown with labelled differences, and then excluded from the locality, what wonder if a helpless eunuch is found in the trade of flesh? Take into the consideration the impotency and immature genitals and think of the other members his hungry guilt has! Let him bury his shield of vulgar words, can anyone guarantee his protection from assaults of gender and that he will not be sneered at? Think world and think again, when a eunuch gets into such facts, what would be the cleanliness of the society in which absolutely males and absolutely females are jumping into the trades of filth; jumping with all their might, from various heights.

Our condition was critical; we decided that we would vacate the room today. Fortunately, while we were leaving the police station, Sharma ji appeared, I requested, "Sir, we want to vacate the room till 11 AM today, can we contact you in case any problem arises?"

"Why not, take my number, call me and I will get there within few minutes, don't fear during shifting," Sharma ji replied.

The soft sound of solace is as mighty as a surging sea wave. We adopted a calm attitude during shift and landlord quietly received the money we offered. After a month in the new room, Dharmendra joined us. Considering his urgency of having a job, we got focused on him. Enveloped in the waves of affection, filled with the waves of gratitude, Dharmendra would whimper at midnight covertly in his blanket, oblivious of the fact that I might be awake.

Clad in stars
With moonlit face
The dancing night leaves
How the stars appear!
Distance-born closeness
Fastens
Traversing the shabby coverlet, the moon
Sojourns on an eye
Eyes
Whether big or small
Inherently know
To open, to close
To rest in evenness

Deepak Dwivedi

In the midstream of responsibilities, Dharmendra was working very hard for our hopes. After three months of incessant efforts, he got placement with a good salary package in a company at Paonta Sahib (Himachal Pradesh). At the time of leaving, Dharmendra's gait assured of a long journey, his soundless throat, wordless mouth said everything, his eyes gave a glimpse of the rain-reservoirs and his gestures manifested a phase; Dharmendra's heart had crowned my affection.

After Dharmendra's departure, Rakesh also got an opening in a pharmaceutical company, he shifted nearby his office. Ramgopal and I continued studying, to us, nothing was as charming as research. Ramgopal visited JNU every now and then. One day, I received a call from the AIIMS mentor, I reached his office.

"May I come in sir?" I asked

"Yes, come Deepak, come, I was thinking of you," he said.

"See, I need a student like you, I have a job for you here," he added.

"Thank you sir," I said.

"It's a temporary job and is a part of a research project, you fill-up a form that I give you and come to me the day after tomorrow," he explained.

"Ok sir, sir . . . I have to appear in National Eligibility Test (NET), can I come after 3 days?" I asked.

"Then no joining will occur," he promptly replied with a sarcastic smile.

"OK Sir, I come the day after tomorrow," I replied.

"No, no, you have to stay in the lab today and tomorrow too," he said.

"O . . . K sir," I replied in a low voice, and left the office.

92

In the lab, a boy aged around 30, began to inquire me. "Who told you about this laboratory? I mean, who recommended you for this laboratory?" he asked twitching his eyebrows.

"I had a call from mentor," I replied.

"Call! He knows you?" he surprisingly asked in a surprise.

"Yes, he has been my M. Sc. Dissertation guide," I replied.

"Then you must have known him very well, what brought you back?" he punched thoroughly.

"I love research," I replied.

"Is it!" he smirked.

"Have you joined here in any project or any fellowship scheme?" I asked.

"Brother, I am here for Ph.D., mentor knows me very well since long back; so long that you would be a little boy at that time. I couldn't count how many times he called me up for Ph. D., and of course I couldn't refuse although I was already having a very good position in corporate sector," he said jerking his neck many times. Hearing all this, a girl working nearby shifted away with the complete set-up she was working with.

"Let's go for a tea," he spoke looking at the girl, in a voice louder than required.

"No sir, thanks, mentor may call me anytime," I said.

"O Come-on! He is not going to call you up for next two hrs. I know that. He was with me a few minutes ago, stand up, your research will soon get finished if you fear like this," he insisted.

"Ok, let's go, but we will get back in 10 minutes," I said.

"Offo! Ok." He began to step jerking his neck.

"By the way, I would suggest that you should try somewhere else to do research, don't compare with me, I have a strong back up," he said.

"I think, if the atmosphere remains the same, I will soon move away," I said.

"Does anyone else know you here?" he asked.

"No," I replied.

"Boss, learn a little buttering and back biting for a successful research, you seem to be very soft," he suggested.

"Research has nothing to do with buttering and back biting, ironically both of them themselves are the matter of research; I would prefer to learn about ethics and law," I instantly reacted.

"You will not be able to grow here with ethics and law," he snapped back.

"I don't think so," I said. He became silent and began to focus on tea.

Next day, he did not appear in the lab, mentor called me up in the afternoon.

"May I come in sir?" I asked.

"You . . . you get lost, I don't need a boy like you," he roared.

"But sir, what happened? Have I done anything wrong?" I asked in a surprise.

"You talk of law with the laboratory persons! You will file a case! You spoil the atmosphere of my lab! Go, leave this place," he flared-up.

"It is true that I had a conversation with a laboratory person yesterday, but it was a healthy discussion on research atmosphere, and I spoke of ethics and law in that context, I don't think that ethics and law are meaningless in research," I said.

"Disappear!" he ordered.

Perhaps he would have pardoned me; perhaps I would have joined on that laboratory post the next day, but I earned a heartfelt detachment from that atmosphere, the place where I learnt not to return.

I did not appear in NET. A series of non-scientific events that took place in a well known science laboratory went through me as a severe shock. I was left with my drooping expectations; the expectations of nobility and tenderness, all genuine, I was a student. I was feeling an overwhelming need of a Guru, my ultimate master; who would show me the goal of my life, who would clear up my blurred vision, whom I would love serving, live serving to, and die serving to.

India is the land of Yogis. Here lived the seers of ultimate truth; even today you may expect such seers. Have a slow walk in to Indian philosophy; you will become a seer of a prospered humanity. You may doubt it at first sight, you may already have a conclusion, you, like me might have seen numerous frauds, religious fakes, non-scientific and blind followers, but one day you will find it reasonable to believe that it is possible that many of the existing life forms of India will go extinct in near future, but it is not possible that this land will ever be devoid of Yogis. Let our scientific exploration get to that knowledge and before that, let me have a declaration that merely every creed and sect of the world puts in a different way—Reach love, return to it.

I began a search for a Guru. My yearning had a deep impact on Ramgopal, he was always with me wherever I went. Whenever we heard about any spiritual giant, we rushed to him, but the same thing happened each time; we would see massive crowds, listen to the giant carefully

Deepak Dwivedi

from beginning to end, and return disappointed analyzing the whole meaning of the event.

After a month full of searching, I turned towards prayers, day and night, praying in every conscious breath, "O Guru, where are you? See I am searching for you here and there! O God, meet me with my Guru, show me the way to him!" My daily routine was disordered, most of the time I would sit or lie praying in the room, sometimes in restlessness, would rush outside praying for Guru in the streets and waiting for him here and there on roads, on subways, in fields, outside the temples, beneath the trees and so and so, and would return to the room muttering "O Guru! Another day is passing by keeping me away from you". Many times I would fall asleep while praying; many other times, would go sleepless for one or two days. Ramgopal would himself finish all the daily chores; making room tidy, marketing, preparing food, washing the utensils, he would wake me up, would make me conscious of my minimum daily requirements and if I refused to eat, he would feed me with his hands, would add everything in it to make it tasty for me, he would lay my prayer mat and clean the place. I was plunging in prayers, Ramgopal soaring up on duties.

My free-falls, your supports
Waking me up, making me move
I know I am substandard
For I respect your reasons, it didn't prove

I have learnt to attend
Your true advice
Yet I fail to comprehend
Why you choose to sacrifice

Descended to persuade
My sulked fate!
For this you ignore
Your life-estate!

They have their own fabric and a loom
With whom
A purpose transcends
To adorn the world with friends

Two months passed in search of Guru. Ramgopal discussed my situation with Rakesh, he would come to us every weekend. They both wanted to give an important suggestion to me but were awaiting a change in my situation; they were serious about the passing time. Ramgopal was thinking about joining a NET study centre. I realized that my situation had badly affected Ramgopal's studies; we both joined a study circle in a nearby colony, Jia Sarai.

One day in the evening, while passing through the market, Ramgopal, Rakesh and I took a stop at a shop. During purchase, the shopkeeper began to talk to me about some new products, but as he shifted his gaze to Rakesh, Rakesh immediately gestured me to leave that place. I was astonished seeing Rakesh's face, it turned pale, his lips blue, and before I could connect to what was happening, he started vomiting and fell on to the ground, we rushed him to the room.

"What was that?" I asked Rakesh.

"I don't know but as that man gazed at me, I felt an intense restlessness," Rakesh replied.

"How are you feeling now?" I asked.

"Better, but it will take a few days to recover completely," he replied.

"You seem to be an easy victim of an evil eye (called *buri nazar* in India)," Ramgopal said.

"Yes, I am susceptible for that," Rakesh replied.

"What kind of consequences do you generally face with?" I asked.

"Physical ailments like fever, body pains etc. My mother tells me that once when I was some three year old, I was sitting at my doorsill. A passer-by Baba (holy man) stopped at my door and spoke to my mother, "Your son is under the effect of an evil eye, does he eat well?" "No he doesn't, he is ill," mother replied. "OK, he will recover and will begin to play in a moment," he said and did something with his broom that he carried and then left, he did not ask for any alms, offerings etc., and then, mother says, I got completely fit and began to play," he replied.

"Various kinds of evil eyes and their different effects are reported worldwide. What do you think? Is there any invariable explanation for this phenomenon? It is one of the major fearing elements in societies; I think it is very important to establish a systematic explanation on this very common issue, isn't it?" I said. "Yes, it is." Rakesh and Ramgopal endorsed.

"I have noticed one more thing. Often, on being stared at by someone whom I am unaware of, I find my sight suddenly falls at the onlooker," I further added.

"Do you feel something different, any sensations etc.?" Ramgopal asked.

"No. I never have a feeling or any sensation on being stared at, however many people report that they feel that somebody is staring at them, I just have a sudden movement of my sight towards the person who is staring, and it is this uninterrupted turn of my neck and eyes specifically at the onlooker which brings a thought to my attention which is "Has the onlooker's gaze pulled my attention? I have also noticed that this experience has always happened within a range of few meters between me and the one whom I catch staring at me," I replied.

"Various things happen in this world, it is all a matter of focus, people generally notice very little," Rakesh added.

"That is all right but if we have an experience again and again, we become simply focused to it," Ramgopal focused the point looking at Rakesh and then, at me.

'Mystery' attracts both, the science and the art. Both the ways of human knowledge revolve around it and during the revolutions both of them pass through each other. On the one hand art takes orientation towards 'beauty' through communications while on the other hand science orients towards 'explanation' through methods. For the former a philosopher has a word 'aesthetics', for the latter he has 'realism', and with these two words he sees two principles clasping each other in a phrase like "Wow! That's a testable prediction".

After a few hours we shifted our attention away from this incidence and the concerning speculations. Rakesh went back at his routine. Again, my urge for a Guru began to push me to the same previous state but this time with some well defined restrictions; the wandering became limited to our way to study circle and with time restrictions, to the area surrounding Munirka. Yet, every night at bed-time, sitting straight with closed eyes on the bed, I would search for my Guru in the realm of brain with the helm of mind, in the hopes of a real clue of him. Imaginations, thoughts, logic-sprints, exhaustive reasoning and eventually, exhausting puzzles would shift me to dreams.

My sleeps, even small naps, like a newborn's sleep are full of dreams and as usually happens with many of us, often minute details of dream objects are left in the memory. 'Dreams' and 'deep sleep' have been a matter of systematic inquiry of ancient thinkers. In Vedic scriptures of India especially in *Upanisads* you will find significant reflections on them. Today we have several classic explanations on sleep, dreams and unconscious states. These explanations are the great works of scientists from disciplines like neuroscience, psychology and physiology, for instance they tell us about eye movements during sleep, the activities of voluntary muscles, blood pressure, brain activity, different sleep stages and time, role of genetic factors in sleep and the relation between sleep and growth.

Ramgopal was attentive to my each and every activity. With his interpretations he often sought from me the logics behind my activities. He and Rakesh had distinct ways on this aspect; Rakesh would never raise any question; he would wait, wait forever for my descriptions while Ramgopal would never wait unless there is no option other than waiting.

But these days I was unable to assemble any logical explanation for my fresh experiences. For some 15 days, I was having dreams concerning Rakesh. The objects and happenings in these dreams were clearer than ever and each dream was proving to be true. Few events turned out to be the replica of what happened in dreams while the others were similar in outcomes. The dreaming happened variably before and after the events took place in reality. Of these 7 or 8 dreams, one was of great importance in which I saw Rakesh wearing a pink coloured shirt, going for an interview and then saying to me on phone "I am selected".

This dream occurred around 6 AM. I phoned Rakesh at 9 AM "I dreamt of you, you were wearing pink coloured shirt, going to appear at an interview, then you said to me "I am selected" on phone," I said. "I am in a bus right now, dressed in a pink shirt, going for the interview, I was thinking to inform you about this interview just after the interview, and you are revealing the result though I am still under pressure but quite happy to know that the result will be declared so early, I call you back after the interview," he said. Few hours later he said, "I am selected."

After few days, another incident occurred to the direction of my dreams and related real events. One day at study circle, we students were having a group discussion on some topics of molecular biology. During the discussion, someone of us asked a girl sitting in the last row, "Do you have any problem?"

"No . . . Mood is slightly different today," she pretended.

"We all are friends, you can share with us the reason behind it," another girl suggested.

"No no, nothing serious, you continue the discussion, I will soon join you". She managed to stay aloof.

Suddenly, Ramgopal whispered something in her ear. After the group discussion was over, she came to me and said, "Ramgopal told me that you know astrology, I have to know something about my future life, could you please help me? Examination is nearing while I am unable to concentrate on my studies."

"I am really sorry for I cannot render any astrological help to you because I have not practiced it since long, I even don't have the literature with me at present," I said.

101

"But Ramgopal never lies, you certainly know something," she insisted.

"Whatever I am passing through these days is not due to my astrological knowledge; it is about dreams which are proving to be similar either in details or in conclusions to the external happenings but I have no knowledge of how it is happening, you take it as I am just an observer, though not a passive one but still find no input of my will behind all this," I clarified my position.

"OK, would you try once for me? Please," she requested.

"OK, then let me try to connect something tonight, you discuss about your problem with me tomorrow," I said.

"Thank you very much," she said while leaving.

"You have tossed me towards an excellent puzzle while you know very well that we both cannot pretend to behave like champions of the things for which we lack a proper comprehension," I said to Ramgopal.

"You always underestimate yourself, Rakesh too says the same thing about you, I know that you can help her, this is up to you how you help". These were Ramgopal's words.

I had begun to think that Ramgopal's actions could never bring anything wrong for me. Now on this occasion I speculated that certainly the upcoming days are going to be very special, but I had nothing on my own, I had only one thing with me—Prayer, and I prayed "O God, if it is appropriate that I should help that girl, bless me with a way to do that". And I went to bed. Around 2 AM, a voice woke me up "Mithlesh!" Ramgopal was sleeping, I concluded that I was having a dream, and then fell asleep again. In the morning my first thought was "Who is Mithlesh?". I asked Ramgopal, "Do you know someone named Mithlesh?" "No," he said. "OK, maybe this name has some connection with that girl, I will confirm it today at study circle," I said.

"Do you know someone named Mithlesh?" I asked the girl.

"Yes, Mithlesh was my paternal aunt, she has expired recently," she replied with a look of surprise.

"But how do you know about this name?" she asked.

"I heard this name in my dream," I replied.

"What else did you see?" she asked.

"I saw nothing, just heard the name," I replied.

"OK, now I believe that you will certainly solve my problem," she said.

"Is it! Tell me the problem however I am as unsure of the outcome as earlier," I said.

Her problem was taking a relationship. I tried to convey her that her problem did not need the astrological and intuitive inputs and that only psychological inputs were needed to solve it. The girl accepted my each and every suggestion with wonderful positivity. After some time she became so confident of her actions and logics behind them that she herself started to solve other's problems. I, on the other hand became indulged in finding logics behind the problems of people asking for my help, and measuring efficacy of very easy and common remedies. I was happy helping people but cheerless seeing the dangerous degrees of their problems, sometimes I would go very emotional and sometimes would fall in 'no-reaction zone' between the great recoveries of some people and terrible tragedies of others. I, for the first time in my life had lived on prayers for such a long time, starting with the search for a Guru to well being of people. Even in a less surety of testable outcomes, I found that praying was not a haphazard and meaningless contribution of religion to any civilization but a systematic and the mightiest thing in human hands.

Ramgopal caught jaundice; I discarded my mobile SIM, purchased a new one and distributed the new number to very limited people. Ramgopal did not like that I focus much on rendering care to him, but he was unable to avoid it. One day he said, "Why do you say people "Ramgopal is a standard example of friendship in modern times?" Because I cannot speak like you, they believe more on you and less on me. One hundred Ramgopals are nothing before you." "For this reason only which is apparent in what you have just said. Ramgopal knows only to help Deepak, it is unbearable to him being helped by Deepak, isn't it?" I replied. He became silent then began to gaze at ceiling fan. Both, a great friend and a great puzzle have the same effect on intellect—they sharpen it.

We hired a cook, a boy aged about 18. Few days later, Ramgopal had a call from his paternal uncle from Dehradun, he left for Dehradun. Each hour of Ramgopal's absence was like a day in desert, he returned after twenty days. The information of his arrival filled me with freshness, but upon meeting I asked him at first, "Why are you so tensed?"

"I will reveal the reason tomorrow" Ramgopal said.

"No, tell me right now," I ordered.

"Paternal uncle says that I must join a job in Dehradun," he replied.

"OK, so when are you going back?" I asked.

"After three days," he replied.

"OK, so you opted to remain tensed for next 12 hours taking this small thing!" I said.

Next day I felt utterly exhausted, fell asleep whole day. When I got up in the evening, I found the food was untouched and the half of the luggage was not there in the room. Ramgopal appeared at around 8 PM.

"Where were you? And where is the room luggage?" I asked.

"This room is costly for a single person, I was searching a room for you, found it in the afternoon, half of your luggage is there, little is left here that we will transfer tomorrow," he replied.

Clasping these words, I jumped into a dimension named Ramgopal. An era with a friend; Ramgopal here, Ramgopal there, everywhere. I found that I am nothing but particles, surrounded by innumerable particles, appearing as Ramgopal, I saw my fragments dancing in his heart and those resting on his feet. Dear reader, If you are really reading all this and if I am really speaking to you then we both know that we are the parts joined through these lines, if this book now wants to slip away from your hands, let it slip away, let us join a dance which is happening here, don't think a little about the dignity of this book, thousands of such books wait to wave their pages around a true man.

"What happened? Where are you dear?" Ramgopal suddenly brought me back.

"See dear, you know very well to understand such situations. Let's go for a tea thereafter I will show you the new room, I hope you will like it," he added.

"Come on!" he insisted.

"OK," I said.

I landed in complete loneliness after Ramgopal's departure. I had phone calls from Rakesh, he was very busy. The cook disappeared without any information. Once again I began to plunge in prayers; now with two purposes: for well-being of people and for finding my Guru. Mostly in day time, I would wander dismally and would get despondent in the night seeing the world rotating on the axis of sorrow. One day in the evening my cook suddenly appeared.

"Where were you for so long? Without any information! I thought you will never return, may be you are no longer interested in working here but this is not a proper way of quitting, you will never progress with this attitude," I roared.

"Sir, I know that I have done wrong and I accept my fault, even being so, I have come to you. I have come to inform you that please don't wait for me; I have respect for one and only person in my life—you. I tried to come here but couldn't come. I need your blessings." While saying all this, he sat down near my feet.

"Oh! Get up, have this chair, tell me about your problem," I asked patting on his shoulders.

"Sir, I am serving as a cook at a very dangerous place. They are body-builders with long stature. They are five in number. They do wrestling. They bring huge money every day. They intimidate people and pose threats on phone. Nobody dares to fight with them, they are familiar with policemen, I am entrapped, I don't want to work there, but they have not paid me for two months so I cannot run away, sometimes they give me 50 or 60 rupees. They slap me whenever they want, sometimes they do it even if I have not made any mistake, just for fun and then they laugh. But this is nothing before the pains of a poor family that neighbours their house. There are four members in that family; husband-wife and two daughters of my age. The body-builders intimidate the family with bad intentions, the husband is very sick and unable to leave the bed. The wife and daughters fear to report this in police station because they know that no one will hear them and they will be killed by those body-builders. One day while they were threatening them, I began to weep, they began to target at me, I begged their pardon repeatedly, then only they allowed me to go," he said.

"Show me the house where these ruffians live," I said putting on a shirt.

"No sir, they live far away, some 20 km away, we will not find a bus to get there right now, you keep away from them, they are very dangerous people," he said.

"Are they more dangerous than death? Have you heard the names of Gandhi, Subhas Chandra Bose and Bhagat Singh?" I roared.

"Sir, I will come tomorrow, I will show you their house," he said.

"First you answer to what I am asking you. Have you heard the names of Gandhi, Subhas Chandra Bose and Bhagat Singh?" I fumed.

"Yes," he answered.

"Do you know that they had no fears of death?" I asked.

"Yes, I know that but I fear," he replied.

"Those who fear will not die?" I began to stare at him.

"Yours problem is that you identify yourself as an inferior person. You have not realized who you are. If you had not have a morality, their doings would have not made you so restless and sad, you too would have managed some 'take it easiness' in all that but this is not with you, and you know that there are people who don't care for what is happening with others, I have also worked with such people, they are highly qualified, they have money, power and people call them 'dignified' but to me, they are not worthy of having a place in the dust of your feet, and you see if I tell them this fact, they will scratch my whole body with their nail-saws," I said.

"There are two ways to save that family," I said.

"What are those, sir?" he asked.

"See you are about 18 and I am 21, now first you tell me whether there live the boys of our age-group in that colony?" I asked.

"Yes," he replied.

"There must be the gatherings of those boys around any nearby tea stall or any talkies," I said.

"You tell them all this, I am truly speaking, they will unearth the ruffians with the unity and cleverness in less than two days," I said.

"You are right sir, and the other way? You said that there are two ways," he asked.

"Do you like either of the daughters?" I asked. He began to hesitate.

"Tell tell, tell me the truth, speak from your heart," I insisted

"Yes, the younger one," he replied.

"That is all right but which one likes you?" I asked.

"The same girl," he replied.

"How do you know that she likes you?" I asked.

"She often says to me that I should give up working there and says that I am dwelling upon the danger,"

"This means she loves you but you only like her?" I said.

"No no, I also love her," he said, his teeth got exposed.

"Then go on and marry her, bring your parents and relatives, the ruffians must know that their neighbours are not alone, take your to-be father-in law to any government hospital, it won't cost much, take this money and work hard for them, you are at the age of diligence and achievements," I said offering some money.

"You are right sir, I will manage it all, yours blessings were necessary, I had not come here for money," with this brave refusal of money, he never came to me again.

It is the result of our own greediness that violence and sex have come so close to each other. What is the relation between these two? Hope that behavioural scientists, neurologists, psychologists, molecular biologists and cognitive scientists will soon be able to explain the factors that give rise to this blemish. What more is needed to extend our pleasures? Let us see how far we are going to propel a purpose called reproduction! See how efficiently we are balancing it—on the one hand, develop new ways to quench the thirst after triggering it and on the other, develop the banks of sperms and eggs! The stimulated hunger nurtured with the spicy foods can never make us healthy; it is a perfect trick of fooling ourselves before the mirrors of satisfaction.

Are dedication and self-sacrifice psycho-activities? Does a dedicated one who is ready to sacrifice himself for the sake of others lacks the balance of logic? A pious intellect, by its bright nature, will convert these questions into immortal waves. He is not ignorant of the importance of his life; he knows the essence of his existence.

Few days later on an evening Rakesh visited me with his new friend with whom he shared a house in Noida (UP).

"Deepak, he is Rajesh," Rakesh said.

"Hello," I greeted.

"Jai Maata Di!" he hailed.

The word implies 'salutations to mother (goddess Durga; a warrior aspect of supreme mother)'

"Jai Maata Di!" I repeated. Rakesh began to read a newspaper.

"I have heard a lot about you from Rakesh, I felt an urgent need to meet with you so I insisted him to get me here," Rajesh said with a gentle smile.

"Yes yes, you must have heard a lot about me, he has a precise account of my wickedness, this certainly has insisted you to seek my whereabouts," I chuckled.

"It is fascinating to know that there is someone who loves to pray," he said.

"Both, the striving and praying are very powerful; to pray is very simple but it is not easy to understand what praying is which is why people with a surface view often fail to recognize the essence of it, I also lack a precise understanding of it," I said.

"What to do with concept of praying while I am not free from the joy of prayers?" he said.

"This means that the descriptions of gods and goddesses from scriptures fascinate you?" I reasoned.

"Completely, and I want to live with this joy forever," he confirmed.

"I think we should leave now, it is too late, we have to reach Noida," Rakesh said.

"Oh yes, let's go," Rajesh said.

"I will soon take you with me, why do you live here in Munirka? Join us at Noida," Rajesh said while leaving.

"OK, I will soon decide on this," I replied.

Rajesh appeared with an empty bag next morning. Before I could understand anything, he filled the bag with my clothes, informed me that I had only five minutes to get ready to leave for Noida, and began to wait for me at the door. We reached Noida. Rajesh and Rakesh lived in adjoining rooms; left side to Rajesh's room was another room. Altogether, there lived 5 boys in the house. The rooms were theoretically divided but practically there was no any allocation of rooms among the boys. All the boys were very simple natured, all were, like me, from small cities. I had become an inseparable part of this family long before reaching here.

Mostly, Rajesh himself would prepare the food for the housemates. It was not because only he knew how to cook, it was due to his passion to render his services and great care to each and every animal. His dedication compelled us to accomplish two tasks: waiting for delicious foods and eating. Rajesh had done Masters in computer application and was working in a software development group in Delhi. His daily routine was thus: To rise at 5:00 AM and then to go for an altruistic walk, he would return to the house after feeding vegetables to cows by 6:00 AM. In next 15 minutes he would do three things; putting grains for birds on the roof, preparing the tea and waking up the housemates to have it. After the tea, he would go for a bath. At 7:00, he would sit for puja which would be over by 8:15. Next he would prepare breakfast and lunch for the housemates, then would leave for his office at 9:00. He would return from office around 7:00 PM and would sit for evening puja which would be over by 8:15 PM, and then he would prepare tea, the housemates would return by then. Further he would prepare dinner which the housemates would finish before 10:00. Rajesh would then sit immersed in devotional songs playing on his computer; he would go to bed around 12:00.

I stayed with Rajesh for a month. I spoke less during most of this period and listened to him. He took each and every task a divine duty, filled with ecstasy, he seemed to me worshiping all the time. He was so affectionate that my quiet presence beside him was deeply felt and interpreted by him. He would appear to be the most delighted person on the earth while narrating the beauty of characters of Lord Krishna and Goddess Rādhā, Lord Rama and Sītā along with Hanuman, Lord Shiva and Pārvatī. It was delightful to imagine the degree of his joy seeing the tears flowing from his closed eyes, in the rhythm of divine concerns and the songs of devotion.

Often
Simple becomes complex
Experiences and explanations
Perplex
In the brainy world
Two observations do meet
At a point to amaze
Having lost enough in the ways

Few days later in July 2004, I received an interview call letter from Mathura, the birth place of Lord Krishna. The letter was from Central Institute for Research on Goats (CIRG), Makhdoom (Farah) for the position of a Senior Research Fellow (SRF). I even forgot that I had applied for this post. One of my M. Sc. Classmates had recently joined here as a Ph.D. student. I phoned him, he said, "You can come to explore your chances, though girls will be preferred". With little hopes of selection and high hopes of finding a Guru, I set out for Mathura.

5

Knots

Goat is great. Goat is cute and intelligent. Goat has attracted man since ancient times. The relations between modern goat and man have improved much as man has learned to care it. Goat's life is worthwhile in both the ways: while living it produces milk and afterwards, its meat, skin, hair and other body parts are of different uses. It is beyond doubt that goats generally don't mind the misbehaviour of their caretakers but the caretakers should not ignore a well known fact—that goats swiftly revert wild.

CIRG functions under the aegis of Indian Council of Agricultural Research (ICAR). It is located on the bank of river Yamuna, in village Makhdoom which is 2 km interior from Farah town situated at National Highway-2 between Agra and Mathura. I stayed for one more day in the institute hostel with my friend who was pursuing his Ph.D. here. The next afternoon, as I came to know that a girl was selected for that post, I left to catch the bus for Delhi. The highway was blocked due to the heavy traffic jam. After a two hours' wait, I decided to leave next day and got back to the institute.

During dinner at guest house, a girl who was carrying out her M. Sc. Dissertation here, extended like always, a meal-full plate to her senior who was pursuing his Ph.D. here and said, "Do you know sir, a new girl will soon join us in our hostel, she is the new SRF, don't you think that the guest is lingering around?" her comment hurt me, but a few seconds later I thought that the girl needs to be mature, therefore I did not find it necessary to reveal her how uncomfortable I was feeling during this overstay!

I left for the highway next morning. After some 15 minutes walk, I saw a hermit coming. I began to think of meeting with my Guru. The hermit was barefoot. As we approached nearer, I touched his feet, he hailed, "Jai Bholenath! (Salutations to Lord Shiva!; Lord shiva is known as Bholenath due to his guileless innocence; he is said to get pleased with anyone who has even a little faith in him, and helps without condition to any one, he does not categorize anyone as good or bad, he loves simple living)"

"Are you going somewhere out?" He asked.

"I am a guest here, I got here for an interview, going back," I replied.

"You have forgotten something. You have to live here for years to accomplish something, Jai Bholenath!" He hailed and moved on his way.

I tried to remember what I had forgotten and opened my bag. The hermit was right; I had forgotten to bring my mobile phone charger and a few clothes. I returned to the hostel once again pondering at the hermit's point "He voluntarily spoke of two things, one of them has proven true, the other should also be true". My friend had left for the laboratory; I phoned him and waited for the key of his room. He came with a colleague of him, the colleague suggested him, "Dr. Sanjay Barua is a very good scientist and a humble person as well, few months ago a boy had applied for Ph.D. under his guidance, but he got a job somewhere and didn't return, why don't you take Deepak to him? I hope he will not refuse to guide him for his Ph.D."

"Research is my passion, I don't want to do anything else, it will be good if I get an opportunity here, but if not here, elsewhere," I said.

"OK, I will take you to him after lunch but I will not talk to him about the purpose of meeting, you yourself will have to discuss everything with him, I will just introduce you to him saying that you are here on a visit," my friend said.

"OK," I said.

On meeting with Dr. Barua, it seemed to me that I was never an unknown boy for him, it was the simplest first meeting that I had ever had with any scientist. He knew about my qualifications, my work that I had done at AIIMS and my capabilities in such an easy way that I began to think on my return "How much better would it have been if I had come here for my M.Sc. Dissertation instead of going to the country's leading institute AIIMS! The mentor from the big and advanced AIIMS had made me dangerously ill with his attitude. Several laboratories of AIIMS might have produced several excellent researchers that year, several important facts might have got established by AIIMS that year, but who knows how many students that mentor made ill that year? God forbid AIIMS has few more like him, it will be a betraying attack on the curing heart of AIIMS".

Think of a child who seems to be growing in search of a fact. The fact that could define and preserve the meanings of humanity. The seekers of facts or say 'factseekers' are identified by several different names; scientists, philosophers, artists, social workers, yogis and so and so on. These different names represent different ways of pursuit. But why do people think that the ways of pursuit are opposite or contradictory? And why do they take it for granted that the differences of ways create ever increasing distances? But let them proceed this way; let us see where they reach. The corner where I am standing can clearly show us that they are all alike in an error; they are assuming that humanity is a crossroad. Humanity is certainly not a crossroad; it is the ground where ways grow. It takes a lot to develop a human, how better would it be to each and every human to always remember that he is not less than a human and that he is living among humans!

Dr. Barua had done MVSc (Master of veterinary Science) and Ph.D. in Veterinary Virology at Indian Veterinary Research Institute (IVRI), Izatnagar. Virology is the study of viruses, and is a branch of Microbiology. A virus is an infectious agent having a protein coat and a single type of nucleic acid (DNA or RNA), lacking independent metabolism, and reproducing only within living host cells. At CIRG, Dr. Barua was working on viral diseases of goats.

He told me to come next day. Working with viruses is challenging, that's why to have research work on them is often interesting to a microbiology student, whole night I gazed at a knot of hope. Next day he forwarded my Ph.D. application for Director's consideration. By the evening, the Director accepted the application and marked it to Education and Research Coordination cell (ERC cell) where I was told to wait for few days for the *office order*, however, the hostel facility was provided the same day.

With a 99% security of approval, I brought my bag and baggage from Delhi, arranged it well in my room and left for my home for next few days. Mostly, Indian parents pass their lives worrying about their children's future. You select any class; any area of India, if you ask Indian parents what they expect from their children, the most common reply that you will have is "That the children get well-settled". To seek love and care from the children's side is not a matter of grave concern from parent's side, at any age, at any phase of life, and I am quite sure that the mightiest relation in most Indian families is between the parents and the progenies, and this is certainly because of extremely caring nature of parents; parenthood is the axle of Indian families. This point, though it seems, is not so obvious, I believe that by conducting surveys we may systematically know about the unrealized facets of family values and the resulting picture will tell us how the families, at different stages, may contribute to love and peace among different societies.

Reaching home, I discussed with my parents about all the unusual incidences that had taken place in my life. Mother said on my dreams that proved true, "You have met with the similar type of happenings that had occurred with your maternal grandmother and similarly with me when I was at your present age. Most of my dreams prove true, telephones have become common in recent times but there was a time when your grandmother and I had the only quickest and quite reliable means of getting the major information about each other and that was dreaming."

On the other hand, on my quest for a Guru and my meeting with the hermit, father said, "After I passed my middle (intermediate) exams, I got majorly involved in two activities; working as a priest in our village temple and teaching the village students. One day in the temple, a surgeon named Dr. Gyan Singh of our village said my father "I wish your son be a doctor, allow your son to go with me, I will teach him". I don't know why he felt such an urgency to grow a doctor in me, due to his teachings, I got selected for Bachelor of Ayurveda, Medicine and Surgery (BAMS) at Banaras Hindu University (BHU), Varanasi (UP). Reaching Varanasi, I first went to a temple; there I began to think "Where would I stay? I have no relatives or acquaintances here." But as I was leaving the temple, a saint asked me, "Child, where do you live?" "I am here first time; I have come from my village to study at BHU," I replied. "Do you have any relatives or acquaintances here?" "No," I replied. "Then where would you stay?" he further asked. "I don't know," I replied. "Behold, there is a room in this temple, stay here until you find a room elsewhere," He said. I lived there for a month. Why the saint spoke to me when there were many devotees around and what insisted both the persons to take a special interest in me, I don't know, certainly such events should be taken very seriously, I would suggest you to carefully observe such events if they are coming your way, stay undisturbed, try to find a meaning in them, such events rarely happen all the time."

After returning at CIRG, I began to wait for the office order. I was having different types of replies from the ERC office "Come after two days" "Come next week" "Department head is out of station, he will join here on such-such date" "Permission is certain to be granted, design your work", two months passed. But it did not worry me because I was sure that permission would be granted. 'Delay' is an inseparable element of most of the organizations working in Indian government system. I am certainly going to avoid peeping into the dynamics of delay as it will itself produce a massive delay. If you have time, have a random walk into said system, you will have an opportunity to see different types of 'Delay-watches' and 'arbitrarily-curved channels'.

CIRG campus was geographically ideal for my upcoming activities. The institute owns a wealth of forest area and is on the bank of river Yamuna. You will reach the river by moving 1.5 km straight after entering through the main entrance of the Institute. The banks of rivers are very important, from several viewpoints; they are the strongest witnesses to the development of life and similarly, to the different ways of life. The Indus Valley civilization, ancient China, Mesopotamia and ancient Egypt have seen the importance of rivers in their emergence. I was chiefly concerned with the geography of the campus due to my growing hope that I might find here my Guru as I was aware of the fact that the forests and the riverbanks are the two most visited places by hermits and saints.

It was not possible to start the work in the laboratory because the official permission was still awaited. However, I was in regular discussion with Dr. Barua on different conceptual and technical aspects of microbiology, immunology and molecular biology, consequently the work was also designed. He was in discussion with different persons taking my office order but he too was getting similar responses as me; he had joined here two years ago. For next three months, I was replied "few new applications have been received; all will be pooled and processed at once, the permission is assured, have a little patience".

I used to go to river every evening in search of my
Guru. On the peaceful way, I would sing the Sanskrit
hymns. The rhythm of singing and the rhythm of walking
got harmonized in a definite pattern; every day the same
distance travelled as a particular hymn reached. I had
learnt these hymns in my early childhood from my
Grandfather at my home, and since I studied up to class 8th
in Saraswati Vidya Mandir, I could practice them for long.
On both the places I had heard the songs of greatness.
Hearing the endeavours of the past greats would fill me
with unparalleled joy, I was looking for such a Guru for
me. I would nip all the rising doubts with a powerful
thought "Why to think that the conditions have profoundly
changed with time? It is the most modern century in which
a great devotee like my grandmother lives, the century in
which a great teacher like Sri Gorelal Tripathi can call a
B. Sc. Student like me to teach astrology, the century in
which a worthless boy like me can have so innocent
parents, the century in which the great friends like
Ramgopal, Rakesh and Rajesh live, why should I assume
that there remains no great in the form of a Guru?"

The area of riverbank where you reach through CIRG
gets divided into two parts in the rainy season; a 20 metre
wide narrow stream is formed some 150 metre before the
mainstream. Therefore, to get to the mainstream, you will
have to cross the waist-high water. From the mainstream
bank, you will find a boat to go across, I never saw
more than one boat here, and one is enough; due to the
uneasiness of the way and from the institute's security
viewpoint, only 10 to 12 passengers a day travel through
this way. Though, you will not find the goats walking
purposelessly here and there any time in the institute, you
will certainly find here two different organisms roving

anytime, however, not purposelessly—peacocks and snakes. Peacock is the national bird of India.

The institute's guest house was the only place for food for the students. The institute's distance from Farah town and unavailability of conveyance were sufficient to seriously explain why it was important to be highly careful taking food-timings. My condition was often uncertain. But the unanimously best cook of guest house Saadik Bhai always offered his special care to me along with peacocks. Whenever the head caretaker would refuse to give me food because of my late arrival at guest house, Saadik Bhai would gesture me to wait and few minutes later in the absence of head, would give me food saying "I had kept it for you thinking that in case you don't come, peacocks will have an extra diet".

Saadik Bhai is a very humble person. Fixed pleasant face expression, fit body, height around 165 cm, shinning dark complexion, pencil moustache, betel leaves in mouth due to which stained red lips, old-brand durable leather strap watch on wrist, look at him while he winds-up this watch, you will come across the glory of it, and you just ask him "what time is it by this watch?" you would not like to know the time from the guest house clocks, Saadik Bhai would spit even the freshly placed betel leaf to tell you the time. Clad in short Pathani Kurta with folded sleeves, 50 year aged Saadik Bhai didn't seem to be less than a lead hero of a super hit film.

Saadik Bhai had a pang for long. Though, he had received great compliments from each and every person that had visited or had stayed at the guest house, none of the institute authorities had ever bothered to consider him for a permanent position. He would confide to me, "I have always tried to prepare the food with the best of

Deepak Dwivedi

my efforts, and people have always complimented me, but I could never ask anyone to promote me as a permanent employee, you may say that this is due to my hesitating nature, or you may say that I have a constant satisfaction which asks to me that is it not sufficient that I got this much?" When does a pang stay silent in heart as it finds someone dear? I would often try to cajole this permanent pang in the temporary service of Saadik Bhai by saying "Saadik Bhai! Frankly speaking, even if you were a permanent employee, you could not prepare as delicious as you have prepared today"

I never went to the river empty-handed; Saadik Bhai would give me flour to feed fish which I would knead on the bank to throw in small pieces into the river. Seeing the fish would hover around the birds, I would take the grains for them. One day, in emotion, I asked Saadik Bhai, "Saadik Bhai! Why do people fight in the name of their individual religions? You are a Muslim, I am a Hindu, will our affection strangle our individual religious faiths or the other way round? Are individual religions dangerous to one another? I often prefer to wander for my Guru and you always prefer to preserve food for me, your religion is known to you and mine is known to me, and we both know that those who put their best efforts to drag the temples, mosques, churches, and so and so in their ill motives often roar to get considered as wise!" "Much can be done in the name of religion, those who know only fighting will fight only now taking this now taking that, and those who love peace, will live peacefully whether they land in religion or in trouble, come tomorrow afternoon, I will show you two peacocks whom I feed daily, their natures are very different, they come here every afternoon, around 2:30-3:00," Saadik Bhai said. "OK," I said.

I reached guest house at 2:30 next afternoon. After some 20 minutes or so, the peacocks began to walk before the window of guest house kitchen.

"Behold! One of the two is walking at 7 to 8 feet away from here; it expects the food at that point, it is calm-natured," Saadik Bhai said.

"OK and what does the other do?" I asked.

"The other walks on that boundary you see some 15 feet away from here, I throw the food to it there itself because it is very aggressive and tends to fight with the former. If I give the food first to the polite one, this rascal will certainly grab its food, but if this rascal is given first, the former will just wait for its turn" Saadik Bhai replied.

"They both appear to be similar in size and weight," I said.

"Yes, I will not let their weight reduce, they both are dear to me but I take more care of the polite one, this rascal's nature is a matter of little worry, however, its nature helps him find the food early. The polite one likes a variety of foods, it accepts whatever it is offered with," Saadik Bhai said.

"These peacocks appear dancing in your heart! Much difference has gradually developed between the human behaviour and animal behaviour. The animal is innocent, it lacks a system of self-awareness due to which it is unable to think of what it is doing, it is just doing and doing, but in man developed "How", in him developed "Why", man then saw beauty in innocence, he valued it, and in turn, the innocence helped him a lot, but a point of worry is that man, with a massive increase in his population is least bothered about the rapid loss of his innocence," I said.

"You are right, the conditions have changed rapidly" Saadik Bhai said.

Have you ever had an enthralling sensation in helping someone even if you have lost a significant amount of energy and time in doing so? Have you ever extracted a feeling of great satisfaction in keeping yourself less-valued than the others? But have you ever been caught into a moral dilemma while sneaking away from your duties and when being cheated? What is humility? Why should one be humble while one knows that on one side even a beast is fairly successful in fulfilling the needs of its life, and on the other side the Nature offers no special gifts to the humble persons? Is humility our natural element or is it an invention of our refined intellect?

Humility is self-educated innocence. Innocence is a congenital trait of intelligent life. Innocence and Intelligence both are attractive, they usually don't get to stay unchanged and unified for long in a human brain, but if they do, a luminous great appears. How enthralling it is to see an innocent childhood metamorphosing into a humble and intelligent youth! Believe me when you will walk through the aisle between the limitation and potency of life, humility will touch its vigour in you, and at that very moment, your life will find an ideal form of it, from that moment you will not hesitate to become the fertile soil for a seed to make it grow even if you knew that it lacks much to succeed in producing scented flowers and sweet fruits; from that moment you will stand unaffected by the efforts of cheating made by anyone, by anyone dear to you; from that moment you will stop cursing Nature for its cruel changes; that moment will take you in an era where you will peacefully meditate, away from loss-profit calculations, away from selfish enterprises, away from the noises of ego, great is humanity for it endeavoured to find a simple charioteer which it lovingly calls humility!

My repeated exposures with the mentalities of the people of ERC office saddened me. Six months had gone hanging around the office. Now, the office was not in the mood to process my file before the joining of new Director. "Nothing can be done before the joining of new Director, but you seem to be a wonderful guy, you have been a meritorious student, you have worked at AIIMS, why are you yearning for a Ph.D. in this small institute? Are you in love-affair with any girl here? What is the actual purpose of stay?" the people from office would often repeat these remarks. "Sir, it is rather tough to find a guide like Dr. Barua, I have seen the heart of this institute, top-class research can be done here, there is no dearth of funds, the institute needs some research students arrive here for long-term research," replying with this viewpoint, I would often forget of the pricking sensations produced by their remarks, however, later on these pricks would badger me.

It was the second month since the joining of new Director and the ninth of my periodic movement around ERC office. One day in the evening on my exit from the laboratory complex, a security guard, aged around 60, asked me, "Are you doing any course here? I was thinking to talk to you for few weeks but I didn't find it appropriate to interrupt you."

"No sir, though I am here to do so. I am waiting for the official permission to start my work," I replied.

"You seem to me different from other students that study here, do you go river in the evening?" He asked.

"Well, there are reasons for why you find me different from others. They have their well-defined routine targets, moreover I usually think something serious while walking, therefore my pace and body language are apparently different, I go river every evening, it gives me hope and satisfaction," I replied.

"Your good name?" He asked.

"Deepak Kumar Dwivedi," I replied.

"So you are a Brahmin, you must be having faith in prayers?" He asked.

"As per the Hindu-caste system, I belong to this class, but certainly I have not found myself to be the knower of Brahman, The Brahman as described in Vedanta," I replied.

"It is nice talking with you; I just remember my past days. In 1971 while at your age, I took part in India-Pakistan war. A few years later, I was appointed at a temple; for next 30 years. I turned towards devotion, so much as my life saw the power of peace, I knew the other side of the fact that nothing was better than living as a true serviceman to mankind, to animals, then, after several years of practice, I knew what devotion and love could actually do, it is really fantastic to meet with someone who is on the same path, God bless you with a great life," saying all this he became emotional.

"Great! Why didn't this meeting take place earlier? Where do you live? Can I come to you daily?" I asked.

"Sure! I live in the building behind the dispensary. I do duty at this gate twice a week, one day at the main entrance and the remaining time, in the forest. If I am not available at these points, you can come to my quarter, and in case I am not even there, you can know my whereabouts from any guard of the institute, they call me 'Shastri Ji', I will pre-inform them about you," He replied.

"OK sir," I said touching his feet and went straight to guest house to inform Saadik Bhai about this meeting, but he said that he had never met with Shastri Ji.

I got familiar with all the security guards within a few days. Whenever, whoever of them I met with, talked about Shastri Ji only, owing to these discussions, I could learn fast about the greatness of Shastri Ji, he was revered by the entire security staff, except the officers who had no idea of his nature and individual life. He had a thorough knowledge of Vedic texts and had a keen interest in astrology. Promoted by him, I began to revise the principles of astrology that my astrology Guru Ji had taught me. It is merely impossible to live a lonely life in a hostel, yet I developed a simple formula of living with the rest of the boys—neither they would stay with me longer than half an hour nor would I with them. But most of the boys had a serious objection to one of my activities which was that I, even in their presence, greeted a security guard named Shastri Ji and a cook named Saadik Bhai with touching their feet. I conceded that from a viewpoint their objection was logical yet from several other viewpoints it was a matter of personal interest, therefore, I had an objection to the seriousness of their objection. Underneath my objection developed a mighty thought in me—I wait, someday, a wise exclaims with pride "A security guard or a cook is my inspiration".

At the corner of the hostel building was a rudimentary canteen; structurally and functionally similar to a spacious tea shop. The people coming to canteen often walked through the hostel's entrance, scientists rarely visited here. One day in the evening Dr. Barua appeared at hostel. Unaware of his presence, I was practicing a tune on flute. As I suddenly saw him at my door which was open, I tossed the flute aside and got up.

"Good evening sir!" I greeted in surprise.

"Don't stop, carry on playing, the tune pulled me here, I got curious to see the artist," he said.

"Oh! I was just trying to play something, you are welcome sir, please have a seat," I said. He came in and sat on the bed, I began to arrange the papers on some of which I had drawn some astrological chakras and on some I had just begun to write a poem.

"What is all this?" He asked.

"It's just some practice work in my interest area, and sometimes arise few thoughts that I enjoy writing on paper in the poetic form," I replied.

"I am amazed to see your activities, every time I find you doing something different, what else is there inside you? I have seen you walking in the fields in the sunny mid days, on the way to river in the evening, and on other occasions, around temples. Are you searching for something? What are you searching for? What do you actually want?" Questions rushed towards me from him.

"I want to know the purpose of my life, and for this I am wandering for my Guru who will show me the truth," I answered.

"Then, why all these activities?" He asked.

"Everything is important to me, and I don't want to take anything as meaningless or useless, anything good or bad, amazing or boring, frustrating or inspiring. I never wanted to learn astrology yet I was lovingly welcomed by a great man, I had learnt to value and respect people from my family and school, and I behaved accordingly. Now I know that I know very less and for this reason I take care," I replied.

"Great! What inspired you to play the flute, to write poems etc?" He asked.

"I take my dreams very specially because they have taught me much, they have supported me a lot, moreover, few surprising things like intuitive experiences of future happenings, flashing of true information regarding the person whom I know etc have happened with my dreams. Now I am looking for an appropriate explanation of all this. Few days ago I had a dream in which I was playing flute, this inspired me to purchase and practice it, however, I have no passion for it as yet. I have no knowledge of what inspired me to do poetry; I wrote my first poem in emotion at the beginning of my B.Sc. I had been taking part in stage performances at school since my class 5th. Later, my late M. Sc. period proved to be painfully challenging, which met with new questions and fresh hope," I explained.

"It's fascinating to know all this, and this way! You have excited me, it's really wonderful," he expressed.

"Oh! Sir, you have got too much impressed! If not taken otherwise, I want to quickly inform you that I am a stupid though I am only 99% stupid," I said, he burst into laughter.

"Well, I want to give you the birth details of some of my family members, would you like to make predictions?" He asked.

"Certainly I would, but I am very slow, it may take 2 to 3 days," I replied.

"No problem, take whatever you require; days, weeks or month," he said.

It is more a stuff of serious thinking than pleasing and motivating to know that someone trusts in you. Dr. Barua trusted in me more than that I deserved. He introduced me to a senior scientist Dr. Dinesh Kumar Sharma who was working here in parasitology laboratory. He asked, "Which language do you write in; Hindi or English?" "Sir, I write in Hindi, I am not good at English," I replied. "Recite me any one of your poems," he said, I recited. Then he went on listening one after the other up to a total of four. "Meet me daily in off hours," he said. He introduced me to each and every person in the institute who either was a poet or at least had a stable interest in poetry. Dr. Sharma was a literary man, his poems were simple, he said to me, "Your subjects are fresh, well-presented, word selection is excellent, you brilliantly convey the points but seeing the vigour of your subjects, I fear lest you should drift away from the simplicity of expression; be careful taking simplicity, have you read *Madhushala*?"

"*Madhushala* of Harivansh Rai Bachchan?" I asked.

"Yes," he said.

"No sir, I haven't, but I will read it as early as possible," I replied.

"Do read it; it is a fine example of simplicity," he said.

"Inevitably," I said.

I began to ponder at his point. I had never composed any poem taking the simplicity of expression into consideration. I just put on paper whatever had come to my mind, in a way it sounded me beautiful. For me it was never a matter of much exercising for a simplification because I had no knowledge of highly literary words and methods. So, would it become challenging to me to produce simple thoughts as I grow rich in literary assets? Thanks to Dr. Sharma, a scientist, a parasitologist who made me aware of the simplicity of poetic expression! But what is simplicity? Why a simple expression is preferable for human understanding?

Simplicity is the best possible arrangement of the elements that can provide a meaning. Be it any structure, be it any function, simplicity deals with the efficacy of arrangements of the elements. The word 'best' employed in the above definition denotes 'the efficacy'. Imagine of a thread in which the knots have appeared due to a jumble, what do these knots suggest? These knots indicate a disorder thereby a need to remove them; they are not at least meaningless because they mean something, and if removing the knots can enhance the magnitude of meaning, we say that removing the knots will bring a simplification. Thus simplicity often removes the knots but not always; sometimes such removal is not required, and hence we must always be aware of the other side of the act of simplification while transforming the meaningless to meaningful and that is—the puzzles of orientation.

BADDAB
Meaningless?
Something mirrored?
BAD or DAB?
'Assumed mirrors'
Divide a meaningless
To produce meaningful
But they often create puzzles
'Puzzles of orientation'
Where aged meanings struggle
And nascent words land in trouble
In an unambiguous world

In next 20 days, I went through four books: Two poetry works—*Madhushala (The House of Wine)* written by Harivansh Rai Bachchan, and *Gitanjali (Song Offerings)* written by Rabindranath Tagore and 2 novels— *Raag Darbari (Melody of the Court)* written by Srilal Shukla, and *Gaban (Embezzlement)* written by Premchand. Three of these books were originally written in Hindi while *Gitanjali* was first written in Bengali and then in English by Tagore himself, I read the transliterated version in Hindi; the book was gifted to me by an undergraduate girl named Sonika of Bundelkhand University at the time of my M. Sc completion. She said, "Sir, I heard your poems few months back during youth festivals, this is *Gitanjali,* and this is the answer of a question that I asked myself "What should I give you today for which you will remember me?" Please read it once". Though I had occasionally been in touch with *Gitanjali,* it was only in 2005 when the sequence of events guided me to reach the essence of poetic thoughts; practical, highly meaningful thoughts. Tagore received Nobel Prize in 1913 for *Gitanjali.*

Reading *Madhushala* produced two immediate effects on me: I began to recite the verses of it, every verse of *Madhushala* ends in the word 'Madhushala', and the other effect was I realized that the metaphorical depth in a collection of verses is greatly enhanced by the use of few constant words with different combinations. In *Madhushala* these words are—*madhushala, haala (wine), saaki (server), and pyaala (cup).* If you want to feel the vigour of *madhushala,* hear it from a legendary actor Amitabh Bachchan, son of Harivansh Rai Bachchan, you will have an opportunity to notice how efficiently can an actor understand the limits of acting. A true actor is he who not only possesses the brilliant art of simulation but also has efficiency, courage, and adherence to the reality.

I learnt the relevance of two-pronged effects of satire from *Raag Darbari*. Srilal Shukla was awarded the Sahitya Academy Award, the highest Indian literary award, in 1970 for this novel. *Raag Darbari* presents the picture of a North Indian village that undergoes a downfall due to the polluted motives of powerful political masterminds and of what an educated guest from a city realizes staring at moral loss. The mighty corruption treading on the weak, helpless righteousness goes on to construct a wall ensuring its safe future. *Raag Darbari* depicts a protuberant reality of a polar system in the ink of humour.

While reading *Gaban,* I came face to face with the simplicity of literary accounts of common man's psychology. Premchand is one of the most celebrated writers of India. In *Gaban* he delineates the conflicts of an individual's greed with the social norms in North Indian society in pre-independence India. If a poor, but educated man says "Nobody can understand my reality", he is absolutely wrong; he, after fulfilling his daily needs and duties, must strive to approach to a man who can spare few minutes for him telling what a simplifier like Premchand has left for him, exclusively for him, to his man—a poor man.

There are a few in the world who love everything in you. If parents are not taken into the account, most of us often yearn for even a single person who can afford loving everything in us. But I met with a person in CIRG in whom, I can assert, you will love everything, even if you are quite sure that you are a fairly crabby person. The person's name is exclusively mentioned in next sentence. Ramdas Bharti. People called this human by his full name but I, as Bharti Ji. When I was introduced to him, he was precisely few years old; approximately 55 years old. Since the question of his actual age may make utterly confused to even him, I expedite the issue by disclaiming that the age given above is surely and purely approximate. It is now ascertained by disclaiming that I would not have to face with an innocently serious reaction from Bharti Ji

"This was not expected specially from you, you yourself put my age wrong!" and in a response to his reaction, I would not need to say him "Why to domesticate a juvenile hope at this age Bharti Ji, would you yourself put your correct age at once?"

According to Bharti Ji, he was appointed here on the job of mowing the lawns. Mowing for years, one day he started doing something which he had never planned for; he fell in love with goat, he began to stare at goat several hours a day. He committed much in this love. Dr. Sharma told me, "Our institute occasionally organizes *Kavi Sammelan* (Gathering of Hindi and Urdu poets of India in which participants recite their poetry before an audience), of which I am the organizer. I want you come on the stage. I suggest you to meet with Ramdas Bharti, the member of our technical staff, he is a poet. I am a co-author of a book *Bapu Ki Bakri* (Gandhi's goat) of which he is the main author; the book is a poetry-work. In spite of being comparatively less educated, he had worked hard to collect the information regarding the scientific facts related with goats, their behaviour, reproduction, different nutritional requirements and endeavoured it to present in poetic form in *Bapu Ki Bakri*. Today, on the experimental side, we admire his speed and perfection of blood collection and his efficiency to vaccinate 100 goats at once!"

I left to meet with Bharti Ji. In the way I came to know that he was recently appointed at post-mortem house, I got there and asked a person, "Sir, would you please guide me where do I meet Bharti Ji?"

"Maybe mowing somewhere or maybe jabbing goats elsewhere, but why do you want to meet him?" He asked in reply.

"Oh! Just for an introduction, I have never met him before," I replied.

"His whereabouts are uncertain, do you have a mobile?" he informed and asked.

"Yes, I have," I replied.

"OK, keep it in your pocket itself, don't show it to him, nowadays on seeing mobile he begins to ask "I want to purchase a second-hand mobile, is your mobile on sale?"

"OK, I will take care," I said.

"Try at other places to know his whereabouts, by the way, a *Kavi Sammelan* is going to be organized within next 3 days at the institute, there you will automatically identify him, the man who forgets his poem while reciting it on the stage is Ramdas Bharti," He said.

"Oh my God! It appears that he is your close companion which is why you are telling about him with this level of transparency. You may further strengthen your relationship by giving him a valuable suggestion—while purchasing a second hand item, it is important to find out whether it is a single-handed-second-hand item or a multi-handed-second-hand item," I said.

Towards the evening, I got to meet with Bharti Ji. I had never seen a 55 year old young man like Bharti Ji. He made me sit on the carrier of his bicycle to drop me at hostel and kept on wielding paddles after paddles until he achieved desired speed of the bicycle and his panting. Crossing the midway point he turned his neck back towards me and said, "I have respiratory problem." I requested him again and again that there was no need to perform such an adventurous exercise, on this he said, "Back brakes are not there, they were lost somewhere in the way this morning, front brakes are weak, I don't want

135

to apply them frequently, your hostel is not far from here, an upward slope is nearing, let me concentrate." Finally, he could anyhow avoid falling on the ground while he was trying to stop the bicycle by lifting his leg high to a raised surface.

Man's heart and his eye must be the cleanest parts of his body. Bharti Ji had passed years of working daylong in sun and dust yet neither his heart caught the dirt nor his eyes crept in blurriness. He grew as an idealistic man. He was familiar with several poets frequently appearing at *Kavi Sammelans* at national level. He began to introduce me to different poets, and always tried to explore opportunities where I could share the stage with famous poets. As a poet, on going for my first *Kavi Sammelan* with him, I was prolifically guided by him in the way. Owing to his few revolutionary suggestions over the shortest route to the venue, we both struggled to appear at this *Kavi Sammelan*.

"This route is well-known to me, I am taking you through the best possible short-cut, very few people know about this route, don't doubt, just follow me" he kept on repeating these words until a newly constructed 6 feet boundary appeared right before us, and even on this, he didn't lose his mental balance and climbed up the wall, moreover, began to promote me to follow the same in the name of youth power. I was halfway on the wall when he shouted "Don't climb, there is no benefit here, we have got on to the wrong wall, here on the other side is a wide trench in which someone has left a thorny bush!"

Hanging on the wall with the major help of single hand, I held my forehead into the other palm "Bharti Ji, now please don't get down in haste, stay a little over there and try to visualize whether there is any route nearby," I shouted. "I cannot visualize properly at dusk, the glasses are in the lower pocket, if I now try to take them out, I may or may not break some of my body parts but certainly the glasses, nothing sounds better than getting down, I am leaving" saying this he clasped the wall warmly, wielded the legs for some time, and landed safely. Thereafter, we walked one km back and ran 3 km on the thoroughfare. Seeing him running ahead during this entire run-time, it became apparent to me that Bharti Ji was a time-punctual person.

Though the amount of sugar in blood must be appropriate for the balanced functioning of body yet the sweetness of character does not depend on the sugar consumption. Bharti Ji's sweet character was transparent like this fact but it was valueless for the persons around him. You might be thinking what the motivational status would be of the less populous place like CIRG where lived the contemporaries like Dr. Barua, Saadik Bhai, Shastri Ji, Dr. Sharma and Bharti Ji? The answer will disappoint you. It is rare to see people extracting 'special' from 'general' in time.

Great care should be taken during to and fro transitions between 'general' and 'special'; the adjectives 'general' and 'special' have been found to be associated with a variety of occurrences: general view-special view, general man-special man, general thing-special thing and so and so on. In the honour of my hopes, I have a request to the champions of different fields that is they try at least once again to make a trip to their special world through a general route. I am sure that they will effortlessly understand the seriousness of my hopes for they are born to the loving hopes of 'a man with general viewpoint'. In me that man asks "What are the immediate and sustainable measures to save the human values in a highly educated as well as in a less educated society and how to accelerate the development of morality without burdening Law?" he concedes that morality has its own physiological limits, yet he is sure that there are gaps that lie long before these limits. I have reached *Knots,* here begins an upward slope where a 'man with general viewpoint' comes face-to-face with a 'man with special viewpoint'.

I and my application completed a year of wait for office order. I had passed the year wandering here and there, and my application, lying somewhere in confinement. But now it was the time of its release, the file was opened and the application removed; it lost its meaning, Dr. Barua got transferred to National Research Centre on Equines (NRCE), Hisar (Haryana). I was told to vacate the hostel within a week. A year aged hope with its jerky takeoff left my shoulders droop. The hermit's remark "You have to live here for years to accomplish something" began to flick its dust off. I began to feel an overwhelming need to meet that hermit again to ask him once the current weight of his remark. But where would I find him? But what other than the effort to search him would relieve my restlessness?

I found out that the hermit lived in a temple on the other side of Agra-Mathura highway. I would have confided him my painful feeling had I not found him talking in an unexpected manner inside a shop. Then I got to know that the market shops were quite familiar with him. It sounded to me meaningless to share with him my predicament for he was no longer the one to whom I had devotion. I paused at a temple during my return, burst into tears and began to pack up my luggage reaching hostel.

Dr. Barua came to me, he said, "You need not go anywhere, I have spoken with a senior scientist Dr. Ashok

Kumar to guide you. He works at medicine laboratory and is about to get promoted as a principle scientist. Presently he is developing the herbal drugs; I think you may carry out a fantastic work with him in that area, he is a very humble person, come with me I introduce you with him, write a fresh application and submit it today itself." "OK sir," I said and followed him.

Dr Ashok said, "I know everything about you, you are now my student, I am forwarding your application, come with the office order, don't be impatient if the procedure takes longer than expected, I mean one or two months." Now the hermit's remark began to rattle my brain once again. Once again it seemed to be right. Though I had lost affection for the hermit, I grew possessive of his words. But all this forced me to seriously think "Is human life a drama or is it a song or something else?" We will have an analytical view of this 'something else aspect' of life in the next chapters, meanwhile, here we need to understand the 'song and essay aspect' of life. Many people say that there are infinite sequels of life through this aspect and many others say that there are infinite tunes of it. This song and essay aspect has deeply touched my life owing to which I have tried to delineate the vigour of it all along.

"Who is it?" I asked
"Wake lest you should say
Why didn't you wake me?"
Said a visiting wave

A formless took the forms
Sound touched the poise of shapes
In that trance, I remember
I was never alone

When my struggles grew
And stars shadowed my fate
In the hut of patience, away from losses and gains,
My pains never left me troubled

O wanderer!
Feet obey the eye's faults, and so hands do
In the dust you travelled through was a flower of path
You slowed, you saw it, never came to pick

She loved, revealed it never
He cherished, murmured it never
They heard some footsteps at their doors
Unlatched them, never latched again

Dr. Barua moved to Hisar, my fresh application
moved to Director's office, Bharti Ji's bicycle moved to
overhauling centre. Though Bharti Ji was in no mood to
go to the overhauling centre but the bicycle was. These
days while returning from the river I was getting to meet
with Bharti Ji returning on foot from the post-mortem
house. On one of these meetings, while I was trying to
explicate him some views of Vedic philosophy, he said, "I
cannot memorise lengthy Sanskrit verses like you, they are
always out of my comprehension. I have studied just one
book for most of my life and that is *Ramcharitmanas*. It
is *Ramcharitmanas* only that gives me words, that teaches
me what poetry is and how beautifully and simply a poem
can flow."

Ramcharitmanas is an epic describing Rama's
deeds. In *Ramcharitmanas,* the 16th-century Indian
poet Goswami Tulsidas, a devotee of Rama, narrates
his master's idealistic deeds in a way that fulfils at least

three purposes: First, to sing a never ending song where everything revolves around the love between a master and a man who aims nothing other than being known as servant of his master. Second, to transform this song in a written account so that this love will heal different pains of the people of all classes, for this reason he prefers an Eastern Hindi language 'Awadhi' over 'Sanskrit' and the third, to nurture the intellect with the art of simile and metaphor and philosophical undertones to the effect that a sad man grows into a happy man.

With or without an open eye, man has always welcomed the knowledge. The four chiefly acknowledged forms of knowledge include religious knowledge, artistic knowledge philosophical knowledge and scientific knowledge. The respect for knowledge brought about a great shift in human understanding. This respect is one important thing and the development of different ways to acquire knowledge is the other important thing. Now, few knowledge forms like religious knowledge are seen to be at the most serious position where on one hand is respect while on the other, criticism. The picture is so frustrating that a man with general viewpoint can't help it. He finds that all the religious beliefs are in a two way struggle: to struggle with the other knowledge forms and to struggle with one another within the kingdom of different sub-forms.

What are the consequences? We know very well, we see that each and every religion teaches to love yet most of these religions often fail to love one another. What lacks? Let us think of respect without wasting a moment so that we may soon discover that it is of little use to keep thinking of respect only; 'self-identity' lies next towards root. The religious development is not a disgrace, the diversification of beliefs is not a crime, it is a natural development, but let us realize that it is extremely rare in religious form of knowledge that one belief unifies with different beliefs without generating a new name for it. I don't know what compels me to still hope in the present conditions that in my lifetime I would get to see a single organized form of world religion that would be valued by all other forms of knowledge.

Ramcharitmanas is a rare example of simplification. Tulsidas was not the first person to narrate the Rama's deeds, centuries before him was a poet named Sage Valmiki who accomplished this in Sanskrit and named the epic *Ramayana*. I have never read *Ramayana* of Valmiki but *Ramcharitmanas* since my early childhood. Two great songs are sung in each and every Hindu family in India—*Ramcharitmanas* and *Bhagavad Gita.*

Two separate events were taking place on the riverbanks these days; on this side I would feed the fish while on the opposite side, fisherman would catch the fish. Since I was not in haste to find out whether or not my act and fisherman's act are contrary to each other, I would return at hostel without any serious thinking on this issue. One day, in the absence of the fisherman, I asked the boatman aged around 30, "Has the fisherman not come today?"

"No he hasn't, few of his fish are yet to be sold," the boatman replied.

"You are appearing here for several months, are you working in Makhdoom Goat farm?" He asked.

"Well, yes but my work hasn't started yet, however, I have learnt enough in past one year," I replied.

"Come, have a trip to the opposite bank, no passenger remains to go across, I am free as of now," he offered.

"OK, but only one round, I have to go far before the dusk, the way is full of snakes," I said.

"OK then, are you going to jump in or should I extend my hand?" He asked.

"Better I take your help, I have not learnt swimming, if I slip, your offer will prove unexpectedly adventurous to both of us," I said.

"Come, then don't you fear to cross the waist-high stream that forms in the rainy season?" He asked while giving a hand.

"I do fear, but because the water is only waist-high, the fear is not felt so much so that I leave the idea of walking across," I replied.

"Well, you tell me one thing, do saints and hermits come here?" I asked.

"Yes such types of people come here frequently but I can understand what you expect, I have not found a true man in any of them," he replied.

"How do you know that no one among them is true?" I questioned.

"They all live in the nearby area, their feats are well-exposed to me, they pretend brilliantly," he said and began to sing the following quatrains of *Ramcharitmanas*.

> *Ehi bidhi jag hari aashrit rahaee. jadapi asaty det dukh ahaee*
> *Jaun sapane sir kaatai koee. binu jaagen n doori dukh hoee*
> *Jaasu kripaan as bhram miti jaaee. girijaa soi kripaal raghuraaee*
> *Aadi ant kou jaasu n paavaa. mati anumaani nigam as gaavaa*

Deepak Dwivedi

> *Binu pad chalai sunai binu kaanaa. kar binu*
> *karam karai bidhi naanaa*
> *Aanan rahit sakal ras bhogee. binu baanee*
> *bakataa bad jogee*
> *Tanu binu paras nayan binu dekhaa. grahai*
> *ghraan binu baas aseshaa*
> *Asi sab bhaanti alaukik karanee. mahimaa*
> *jaasu jaai nahin baranee*
>
> (Bala Kanda 117 (1-4)-[1]

Tulsidas is lovingly addressed as 'Tulsi'. I have tried to transliterate (given below) these quatrains preserving their poetic essence. Although these quatrains have been previously transliterated by several scholars, I have done so for two reasons: First, These lines were sung in a rare joy by the boatman to me, the incidence had several effects on me, therefore, to me, a matter of special interest. Second, I admire Tulsi's adherence and his style of narration. Tulsi takes two names in these quatrains—Hari and Girija. Hari is he who removes the pains and sufferings of people by enabling them see the ultimate truth, is a synonym of Vishnu. Rama, the master of Tulsi, is an incarnation of Vishnu. Girija is a synonym of Parvati. Parvati is the consort of Shiva. Since, according to Tulsi, *Ramcharitmanas* is not only his individual song, it is also sung by Shiva to Parvati, in the line where the name 'Girija' is taken it presents that Shiva is singing these quatrains to her, and Tulsi is writing whatever is happening before him. Further, Tulsi addresses Rama as Lord of Raghus who was the great-grandfather of Rama; the dynasty was famous as the Raghus (descendents of Raghu).

The world though unreal and painful, rests on
Hari, like a man in aches
Beheaded in his dream, not rid of grief until he
wakes
O Girija! The gracious Lord of Raghus
removes such a delusion
His beginning or end is beyond the reach of
reason
Sung by the Vedas, his deduced glory is thus
He walks without feet, hears without ears,
Performs actions of various kinds without
hands
He enjoys all tastes without a mouth
Speaks without voice, he is the Yogi of speech
He touches without a body, sees without eyes
Catches all odours without the organ of smell
His ways are thus beyond the nature
And his glory, beyond description

(Bala Kanda 117 (1-4)-[1]

"Great! I didn't even imagine that this river would someday show me such a beauty. It appears as if this river and this boat taking us are flowing into *Ramcharitmanas*," I said.

"Well said. I am married and have children too, I have had various delicious meals, tasted several types of joy but truly speaking the joy that *Ramcharitmanas* gives me is incomparable," he said.

"It appears that you have a keen interest in literature, what are your qualifications?" I asked.

"Class 5th pass. I know nothing about literature," he replied.

"Ok, Then certainly you have an interest in poems," I supposed.

"Yes I have, but the poems must not be complicated," he confirmed.

"You possess a much higher degree of intellect than that is required for rowing boat," I said while jumping off the boat.

"It is joyous meeting you, tomorrow I will come here little early, we will have some more time for such conversations," he said.

"Same here. I too will get here early from tomorrow onwards," I said while leaving.

I had daily meetings with the boatman for next one month, afterwards, never. A 45 year old man began to appear at that boat, he had no information of my friend boatman. For a month, I studied nothing other than *Ramcharitmanas* and *Bhagavad Gita*. Meanwhile, I was further told to wait for next two months for the office order. I was also informed that the order would be issued only after my selection on the basis of previously obtained marks and performance in interview.

Boatman's remarks were special to me. His remark on the reality of several saintly looking men had a profound effect on me; now, the river area was no longer the area of my interest to search for my Guru. How strangely sensitive the attachment is! It gets injured much more readily than it grows and hurts much faster than it heals. Now I was going to river once a week preferring to wander in the neighbouring villages to find a saint.

By following all the remaining procedures, I received the office order permitting me to carry out Ph.D. work. Seeing this order, this long desired letter, unjustly late by 15 months, I trudged in frustration for several hours. The only thing that defeated my frustration was a powerful thought that glided over my lashing thoughts "It is true that I suffered a lot due to those who derived their joys in laughing at me, harassing me, yet I could bear them because to me a purpose was more important. Though I could not share my pain with anyone nor it is removed yet it is equally true that it has faded. I gained love of Dr. Barua, Saadik Bhai, Shastri Ji, Dr. Sharma, Bharti Ji and the boatman. After all, it is my life. It will be of great worth if I try to pick lessons from my dispersed pain until it is superseded by overall gain." I joined the Medicine Laboratory and started to prepare my research proposal in the line of antibacterial herbal drug development. According to the MOU (memorandum of understanding) of the Institute, I got registered at DR. Bhimrao Ambedkar University, Agra (U.P.).

In a general view we all seem to be familiar with what is research. But on being asked "what is experimental research?" one may need to relook at what he knows as research. The English word 'research' is derived from a French word *recercher* (verb) (noun; *recerche*) which means 'to seek'. One may find a precise definition of research in World English Dictionary as "A systematic investigation to establish facts or principles or to collect information on a subject". But why do we need a systematic investigation? Why to establish facts and principles? What is wrong with us?

And why should we stop questioning? We should not restrict ourselves, we must expand all round with questions, if we can. But why do we have questions at all? The questions are organized, systematic expression of a primitive trait of us that we now know as 'curiosity'. They are the bridges that enhance our understanding on

Deepak Dwivedi

merely every aspect of the universe that we live in. The universe which is full of events and information touches a highly complex and functional human brain incessantly with its all pervading facts, how can such brain lie away from curiosity? A sudden observation on fire led a man to possess an unprecedented power, he realized that fire is power. We now know that knowing is power, knowing is peace, knowing is a highly rewarding event for a brain, we now know the history and scope of a gradual, immutable investigation.

Research can be done in various ways; these ways, in the language of research, are defined as 'methods'. Methods differ because our aims differ. Aims therefore must be clear and specific; these clear and specific aims are termed as 'objectives' in a strict sense. But how would one find an objective and a method? The next concern would be to find a 'clear objective' and an 'appropriate method'.

When a man with a general viewpoint 'observes' the events happening around him, and suddenly, because of his nature of 'wondering' he takes his eye of 'imagination' closer to a particular happening or a thing, he eventually initiates 'associating' the facts with one another and then a bundle of facts with a 'cause' which he thinks is responsible for those happenings or is at least associated with them. He now grows in a man with a special viewpoint; he has a defined "problem' and a 'hypothesis', a testable prediction spanning a considerable length of his 'reasoning', his associating becomes concrete and reliable and his problem valid. He is well set to establish a clear line between the 'cause' and its 'effect'. He starts to see his world in the shades of his problem. His readiness, this striving of him leads him to find an appropriate method.

Modern research has several advantages. It has a great treasure of already defined problems and methods. These problems and methods were defined by great thinkers who lived in our world before us. They gifted us with their great discoveries and inventions before they died. They defined the ways of how to increase our pace to solve the problems; they developed 'techniques' and seeded the world with 'technologies'. But the world is not so simple, it is full of challenges, and the challenges seem to be ever rising, there are countless problems yet to be solved, there are countless meanings yet to be derived; who knows how many principles are yet to be defined when countless facts are yet to be recognized?

There are two broad methods of research. If we have a question like "What may be the cause of a disease that makes people coughing?" we are "aiming" at a particular aspect that we have 'observed'. Here, our question is specific or in other words, our focus is narrow. To find the answer, we need to collect some 'samples' or 'specimens' from the diseased individuals so that we can carry out 'tests' or 'experiments' to validate our testable reasoned guess, our 'hypothesis' related to the disease and its cause. We perform the experiments in the controlled set of conditions in the 'laboratory'. For the validation of our hypothesis we need to take a set of testable materials called 'controls' with which we can compare our test samples. The controls don't vary, they remain constant, but there are things in the test groups that vary. These varying values or characters or categories are called 'variables'. The experimentation generates information on our problem. This information along with all related information is called as 'data'. With this data, we 'analyze' the outcome, often employing statistical methods. This is quantitative research. In addition to the experimental studies, correlational and survey studies more or less follow this method of research.

If we have a question like "What may be the effects of a disease on the lives of people dwelling in a city?" Here, our question is broad; we want to collect all the possible information on the thinking and behaviour of people living in a city, including the patients and healthy ones. We may expect some changes in their conditions, but without determining any outcome in advance. Here, we don't need a laboratory; we conduct our study in a 'natural setting' following a 'standard process'. We perform 'documentation' of the things that we 'observe during our investigation'. This is qualitative research. It seeks to explore the facts that may be further utilized in quantitative research. Social research including historical research, case studies, studies that focus on individual experiences and studies that focus the cultural aspects follow this method of research.

Academic degrees have a special position in the world of research, Ph.D. has its own. I have no special knowledge of the actual philosophy of Ph.D. in different countries but I have carefully acquired the knowledge of the practical picture of this degree in India. Ph.D. in India is the most attractive and dramatic form of research. Ph.D. here is not only a matter of follow ups but also an issue of give ups. Here, more than the serious discussions on different subjects of Ph.D., Ph.D. in itself is a subject of critical discussions. For many people here, Ph.D. is a grand, dark tunnel of knowledge, for others it is an open mine of exhaustion and for many others it is a tale of doubtful intentions. Various Ph. D. scholars, on the basis of their individual experiences, expand it differently like

'Person in Hypertension and Depression', 'Permanent Head Damage', 'Physical Harassment Degree' and so and so on. Whatever comes as an expanded form, Ph.D. in present India is an emotional degree.

In India, Ph.D. students and their guides often have a complicated association; one fourth of the total time it runs tensed, a similar part it progresses weary and the remaining time, stays uncertain, however, on few occasions there are the peaks of care and respect. On my Ph.D. registration, Dr. Ashok said to me, "There reaches a time during Ph.D. when a Ph.D. student asks to himself and to his reliable colleagues many times a day "Has my guide gone mad?". I think when a student becomes sure that the time for abusing the guide has arrived, he should speed-up his thesis writing and if he finds that he has gathered nothing to write as yet, he must speed-up the experimental work." Dr. Ashok's affectionate guidance compensated for my 15 month wait of office order. His humbleness was famous, not only in the scientific and working staff but also among poor men who came to him with their diseased domestic animals. Falling before his sweat nature, even my thoughts preconditioning me for an annoyance to him would fool me changing their shapes.

Becoming a bona fide Ph.D. student had a practical effect on my thinking; my curiosity was now growing as bifurcated. On one branch it was strictly scientific while on the other, rationally religious. In the middle above both the braches was gliding a question "What is the purpose of my life?" This question was not unknown to me; it had been visiting me occasionally since 1999, when I was an undergraduate student. But it was only during my Ph.D. in 2006 that this question became my loving one. This question gave rise to several hundred questions over time

"What is the purpose of life?" "What is the ultimate reality of everything?" "God? Who is he? How is he? Where is he?" Amid these questions unfolded my two quests; to meet with my Guru and to know Nature.

I began to overwhelmingly realize the lack of time, why wouldn't I when I had already spent 22 years of my life living oblivious of my loving question? Thanks to frequently appearing sorrow and fleeting joys that pushed me to look at causes. Owing to scientific education during my adulthood and the religious education during my childhood, I was not completely ignorant of the reality and knew well that in merely every part of the world are found the persons who are overwhelmed by questions.

But even if raised by many or made famous by many or even solved my many, few questions remain personal to us. They touch us so deeply, they travel in us for so long that we don't want to let them change their form in us or their place away from us. Even if solved by one method or in different ways by many people including us, we never ignore them. However, the same fact does not apply to the answers. The question becomes ours as it is, in its original form; the answer is validated again and again and there always remains a probability that the existing answer will be replaced by a better one. It is a widely felt fact that our loving questions produce such a momentum in us that never want to wait for permissions, office orders, funds and fellowships, which exclaims to sacrifice the bodily assets too. My research project, after the evaluation of Institutional research panel, was forwarded for the consideration of the University's research committee. It was May 2006. With the submission of my research proposal, I thought that I must try to understand Nature and then only I would be able to find the appropriate answers for my loving questions.

My weekly routine was more organized than my daily routine. Merely every Sunday I would hover around the neighbouring villages to find my Guru, two days for few hours in the night I would study religious texts, one day in the night would practice astrology while in the day time and few hours in the night I would do my research project work and related studies. I never lived a separate life away from my colleagues; I had enough time to be with them. I had interesting chats with them and they saw an entertainer and an astrologer of practical importance in me, however, my quests for finding a Guru, for finding answers to my questions, therefore, my wandering and overwhelming restlessness sounded weird to them. Such a need caught no one's attention, though they casually raised questions on the validity of my quest, no one stayed to give an ear to my explanation on the problem. Over the months and then years I realized that such quests were not seen by them to be worthy of their attention, they were not required in their lives, as a consequence they occasionally expressed their sympathy and sometimes felt pity for my serious condition.

Six months had passed since I joined the laboratory. In this period, all the research students working before me had completed their works. My study was different from the previously conducted studies in the laboratory; it involved the methods of microbiology, molecular biology and chemistry. The project had two phases: in phase-1, I had to study antibacterial activities of different extracts from 25 different plants against the "drug-resistant" bacteria that caused diseases in goats. I chose 3 diseases of goats; pneumonia, diarrhoea and mastitis (mastitis is characterized by the inflammation and swelling of udders and by the biological, chemical and physical changes in the milk), and in phase-2, I had to find out the chemical groups present in the plant extracts that would show potential antibacterial effects, then to test them mixing in varying combinations of two and three different extracts to check for if there is any enhancement in their activity due to the effect of mixing, or in other words, to evaluate whether they work better alone or in combinations.

We know that we should know about diseases because we need prevention and cure. There are diseases of which we have cure and there are diseases of which we are striving to find a cure. A disease is called 'infectious' when it results from an infection with an 'infectious agent' or a 'pathogen'. Infectious diseases transmit from one person to the others; they are thus transmissible or communicable diseases. Non-infectious diseases or non-communicable diseases are non-transmissible among people, because they are not caused by infectious agents or pathogens, but many of them are genetic and thus hereditary; a child may inherit them from his parents.

Diseases have played havoc with humans, non-human animals and plants. It is this disease that has made struggles for life horrible. I know that every religion of the world has an explanation of deaths and destructions, of such sufferings of individuals and masses but I don't know whether diseases affect gods or not; in any possible way. Perhaps gods, if they are watchful on such havocs are much less emotional than humans, perhaps their norms are profoundly different, perhaps their world is entirely different, and perhaps it is we who selfishly try to see them in our world in any possible way or form so that we have an opportunity to surrender. I am someone with very little knowledge and unable to conclude on whether there is any god for me or not, but for my unbiased reasons, I opt to wait to let him come to us at his will, in a form as real as light, I will wait without wasting a moment of my every possible effort with a clear voice in me "I will never embrace a shadow, nor will I ever welcome an imposter".

Have a look at a growing flower or a young one of an animal, how beautiful, how cute it is! Can you, at first sight, imagine that it is incessantly struggling for its existence, and as a matter of chance, has a good time with all the necessary requirements available for its growth, its beauty to express? We know that we are in the same 'environment' that it is in, we know that many of the living creatures are our 'food' and we are also aware of the fact that we are also desirable 'hosts' for a 'pathogen' that can derive its nutrition from us, and that can multiply within or on us at the cost of our cells, subsequently endangering our survival.

Like in every phenomenon that occurs in Nature, the environment that we share with the other organisms and objects is the central principle in the transmission of diseases. The air, water or anything that travels in our environment can carry inside us and outside from us the materials as well as the tiny organisms. These tiny organisms may be small germs that we can see directly, or even smaller which we cannot see with our unaided eyes. Such minute organisms are called 'microorganisms'; their sizes range in micrometres which is one millionth of a metre and one thousandth of a millimetre. The study of microorganisms is microbiology.

When a small insect crawls up on our body, we can feel the sensation of a movement; we can identify the site of movement and with the help of information available in our memory due to previous similar exposures, we can also imagine that this sensation is crawling sensation of a small insect, not a biting sensation or a pinching sensation. This is done with the help of 'nervous system', which, along with its several functions, is responsible for production and detection of sensations in our body. But what happens when a microorganism invades our cells or moves over them? Is any sensation felt by us? No sensation is felt, our systems have limits-maximum and minimum; there are several levels of reality that we cannot perceive without any external help. But to our pride, most of the external helps that are available to all of us are developed by us. The history of such external helps is golden, it is enthralling, it is magnetic, let me become a little dream-dove, let me expand my small wings and sing . . .

O human, O great human
Your path I love to fly over
The path broaden brighten
O human, O great human

I will perch, I did perch
By the gem first-ever
Gem of faith-religion
O human, O great human

Over and over, I glide over
Love-springs of knowledge
With unaided-reason
O human, O great human

I will fly in a thrust
By the science of endeavour
Reason with invention

O human, O great human
Your path I love to fly over
Your path I love to fly over

Two early endeavours of science, the late 16th and early 17th century inventions that revolutionized our vision were; microscope and telescope. These instruments were the result of the systematic studies on optics, invention of lens and mirror with the help of glass; methodologically developed and arranged lenses, fundamentally similar to those that enhance our vision in our spectacles. Modern telescopes and microscopes have several variations and advancements. Telescope provides a view of remote objects by gathering and focusing light to magnify them. Microscope provides magnification and resolution of close objects. Magnification enlarges the images and resolution distinguishes two closely-spaced small objects as separate entities. Microscope is a basic requirement in microbiology.

Some diseases transmit from animals to humans; they are called 'zoonotic diseases'. Living organisms that are closely related in physiology have a similar risk of being infected and developing diseases by a particular microorganism for example, a pathogenic bacterium. There are several animals that are closely related to us in physiology including the animals that we domesticated, our pets, and unwanted small animals that often compel us to accept that they will share our houses as per their wish. Scientists have 'classified' the life forms that have existed or at present exist on our planet; the science of classification is called 'taxonomy'. Classification is categorization, very important step in systematization, and we know that when we look at nature, we see a 'hierarchy'. The word hierarchy denotes 'the order of levels or ranks'. An important task associated with classification is 'nomenclature'; giving an object and organism a name.

We all have at least two names; a social name and a scientific name, both are very important, the former, given by our family is of social concern and the latter given by scientific community, is of strict scientific concern. Social names are individual names, but presently the scientific names don't distinguish one individual from the other. I am quite sure that I am a modern human, and all modern humans including me have scientific name *Homo sapiens*. This is binomial system of nomenclature because it uses two parts; two words 'Homo' and 'sapiens'. This system of nomenclature was developed by the 18th century biologist Carl Linnaeus (23 May 1707-10 January 1778), who was the first one to systematically classify the plants and animals. His work was published in his famous book *Systema Naturae*. I have not placed the word 'great' before his name because in my view each and every man that has contributed to our knowledge, to our endeavour, to our dedication, is great and I am highly influenced by such men for their works will always remain in the hierarchy of human knowledge, they exist in a simple form, such contributions are the 'active principles' in the sapient knowledge.

Let us imagine what we expect to see when we move towards the earth from the space. From a distance above we see a planet, then on moving down closer we see continents, getting closer and closer we see that each continent has several countries, these countries have several states, the states have cities, and cities have colonies that have houses. Likewise, If we take a top-down view of the hierarchy of biological classification, we see seven main ranks; on top we see 'kingdoms' that cover 'phyla', these phyla cover 'classes', in the same way, gradually, we see 'orders', 'families', 'genus' and then, 'species'. In the name *Homo sapiens* the first word 'Homo' represents the genus, and the second word 'sapiens' represents the species, thus every binomial name is the indicative of where an organism is ultimately placed in the hierarchy of living beings. The organisms are placed in a rank on the basis of 'similarities' they have. The organisms of same 'species' have very much similarities and very little variations in their structural and functional organization. But when, due to environmental or genetic changes, the individuals of a species begin to significantly differ from existing individuals in their structural and functional organization, a new species arises; the process is called 'speciation'. We, the *sapiens*, are not the only humans to exist on this planet, more than 2.4 millions to thousands of years before us lived the humans that differed from our species but they were similar enough to share the genus with us, to be regarded as humans.

Naturally, the above information may give rise to a question in our minds; how do the variations among life forms arise? There are so many life forms on our planet and we can only estimate that their number may be anywhere from 5 million to 100 million, of which we have identified about 2 million. Since we are aware of speciation, a question supported by a guess comes to our notice—had there been a common ancestor of all living beings whose progenies, with time, varied and got well adjusted with their environment? The ever increasing wealth of evidences suggests us to say "Yes, there was a common ancestor and that common ancestor was the simplest of all life forms." Various observations on the variations, struggle for existence, adaptations, reproduction and the origin of species culminated in to the most appropriate conclusions in the 19th century. These conclusions were offered by one of the most influential thinkers of all time, the naturalist, Charles Darwin (12 February 1809-19 April 1882), in his famous and revolutionary theory—'evolution by natural selection'.

Darwin's work was published in 1859 in his book *On the Origin of Species by Means of Natural Selection, or the Preservation of Favoured Races in the Struggle for Life.* Darwin, his theory and his book seem to me to be one of the most influential trios of all time. One may imagine the popularity of the trio by noticing numerous copies of the book, many of them slightly different in their titles, cover pages and page numbers, by noticing hundreds of theories implicating the theory of evolution, and by noticing hundreds of authors applying evolution to the subjects they are working on. It also reflects that a variety of different mindsets loves to be among the intellectuals who look like a champion of evolution, evolutionism, and Darwinism.

I believe in 'influence and admiration', therefore I avoid saying that someone is my ideal and I drew this conclusion because I am deeply influenced by Darwin, Einstein and Vivekananda and at least 100 others and admire them. They did the extraordinary things, for this they influence me and because they did not do the extraordinary things extraordinarily in a sense we know this word as unexpectedly or surprisingly or strangely, instead they always did the things naturally, I admire them.

Though I was taught about Darwin and his theory during my class 11th, and my father had brought me the most recommended books for my whole syllabus, I did not like Darwin and his theory. I liked Lamarck and his theory. Lamarck (1 August 1744-18 December 1829), a naturalist, in his 1809 book *Philosophie Zoologique,* gave an elaborated view of his theory. His theory 'Inheritance of acquired characters' embraced the idea of evolution and he emphasized that characters that an individual acquires during his lifetime reach to his progenies through inheritance. Under the physical conditions of life, the organisms tend to use some of their organs more while some are used less. The organs which are used more and more in a particular way gradually begin to develop and those which are not used begin to disappear, the changes pass on to their progenies and this is how the species gradually change over time. There were two personal reasons why I liked Lamarck and his theory and not Darwin and his theory during my class 11th and 12th studies: first, I studied Lamarck's theory long before that of Darwin's, however they were described in the same chapter; I completed the chapter the day before my class 12th yearly exam. Due to some pressure problems, I could not completely learn Darwin's theory, Lamarck's was easily learnt. The second reason was that I liked the well dressed sketch of Lamarck given in my book and not of Darwin. Though, later my viewpoint got changed profoundly on physical appearances and attractiveness of persons. I am sure it is proper to say that the faces are the most confusing elements in the world of beauty. I began to love Darwin's each and every sketch and photograph the day I happened to see an 1855 quote from him that he had made at the age of 46. The quote, as it is cited on Wikipedia-the free encyclopedia referring to http://darwin-online.org.uk was "if I really have as bad an expression, as my photograph gives me, how I can have one single friend is surprising."

Class 12th onwards, zoology was no longer my subject of study; I studied microbiology. During my M.Sc. studies, the pressure of performance in exams was so high that I studied evolution according to the need of syllabus. It was during my Ph.D. when I was wandering to find the most appropriate answer for my loving question "what is the purpose of my life?" that pushed me to go through the fundamental facts of life by every possible means. I grew less and less concerned with exams and competitions while more and more concerned with the fundamentals of life, to find the answers to my questions. For this reason throughout my Ph.D., I, on one side followed the way of science while on the other, tried to collect and study the Vedic texts and wandered here and there for my Guru.

I reopened 4 of my favourite books that I had studied during my M.Sc. studies; *Lehninger Principles of biochemistry* written *by* David L. Nelson and Michael M. Cox, *Microbiology* written by Lansing M. Prescott, John P. Harley and Donald A. Klein, *Kuby Immunology* written by Richard A. Goldsby, Thomas J. Kindt and Barbara A. Osborne, and *Life-The science of biology* written by William K. Purves, David E. Sadava, Gordon H. Orians and H. Craig Heller. Most of the established facts and principles of life being discussed in this book are beautifully explained in great depth in the above 4 books, all of them have different styles of narration. I recently got to study the 1872 edition of Darwin's *The origin of species by means of natural selection, or the preservation of favoured races in the struggle for life* (London: John Murray. 6th edition; from http://darwin-online.org.uk) and concluded that in a similar way that I realized for

Darwin and his theory, several works of various scientists of different fields and their classic ways of pursuing their objectives would have been missed by me had I not earlier opened the book *Life-The science of biology*.

Darwin's theory and our curiosity must now come face to face with each other. The theory says that new species evolve by natural selection: Nature is full of conditions with incessantly acting forces and laws. It is this Nature that selects who will survive in the struggle for existence and who will not; the fittest survives. Changes occur in the physical conditions as well as in the organisms, and the organisms, for their survival, have to adapt to the change. Governed by this continuous natural selection, the rising variations among organisms are inherited into their progenies and some of them slowly evolve into a new species.

How did Darwin come up with these conclusions? What made him to find such a grand principle that nobody could disprove? His patient observations; long observations, years after years, moving into a wealth of facts, all-round in a well-defined direction with the body of reasoned and testable predictions. For Darwin's efforts, for his great contribution to our knowledge, my intellect welcomes to hear a bold voice from my emotions that it was this Nature that selected Darwin to observe and acknowledge the works of previous greats like Aristotle and Lamarck as well as contemporaries like Wallace, that selected him to wander among the animals and plants of different geographical origin, that selected him to breed the pigeons, to understand the variations under domestication, to board for a 5 year voyage on HMS Beagle to make it one of the most famous ships in history, It was this Nature that mesmerized him with the similar conclusions of different men completely unknown to one another, and that influenced him with *An essay on the principle of population* published by an economist Thomas Malthus. This is how science arrives, thus inherits the glory of science.

But Nature did not make a much needed selection while Darwin was alive and was struggling to understand the principle of inheritance. Every student, every teacher, every researcher, if he slowly walks into the history of biology will certainly make a wish for that historical hour "I wish Darwin had met with Gregor Johann Mendel!" Mendel (July 20, 1822-January 6, 1884) died without any fame; neither he nor his great work which he published in 1866 was noticed throughout the remaining period of 19th century. He was the first one to understand the principle of inheritance. His experiments on pea plants, were beautifully balanced; as mathematical as biological, carefully designed and efficiently interpreted. He is now regarded as the father of genetics, genetics is the science of heredity; how information coded in genes travels through generations.

Today we know that traits are encoded in the genes, they appear if expressed by the genes and thus traits are not inherited directly, genes are inherited. But what are genes? We see that in public domain, especially in newspapers few words have come to be frequently used in the context of the genetic information that we all possess, these words are—Chromosomes, genes, DNA and RNA, and mutations. Things are often simplified when we stay with them, when we observe them, when we move as slowly as possible along them and welcome them into our world as warmly and quickly as possible.

Most fundamentally, the genetic material is DNA (Deoxyribonucleic acid) and in some organisms, RNA (Ribonucleic acid) they both are present in every living organism, in every cell because they both are necessary for the flow of genetic information. We know that the world has a great wealth of atoms and energy and we also know that the events in Nature happen according to the laws of Nature. Atoms, therefore tend to form molecules, some of them give rise to big and still bigger molecules that we call 'macromolecules'. These macromolecules are fundamentally the long chains of their repeating units.

Carbon is the central atom of life; it is the backbone of biological molecules. When a sugar molecule made of five carbons joins a molecule of phosphate and a nitrogenous base, it forms a unit of the genetic material which we name as 'Nucleotide'; the arrangement of these units is the fundamental sequence of life, this arrangement is information, in this sequence are the 'codes of life. The nucleotide units, if repeated over and over, form a long chain called as 'Polynucleotide chain'. RNA is a macromolecule of a single polynucleotide chain while DNA is a double helix; a macromolecule of two polynucleotide chains, both held together with complementary sequences. The difference in the structure and therefore in functions between RNA and DNA are due to the sugar molecule (Deoxyribose in DNA while Ribose in RNA) and a nitrogenous base (Thymine in DNA while Uracil in RNA).

An adult human has several trillion cells with each cell containing 2 metre long DNA, one can imagine what would be the total length of the entire DNA if stretched connecting end to end; approximately thousand km longer than the distance between the earth and the sun!

Likewise, the DNA from a bacterial cell, if completely stretched, would cover the cell 100 times! What a degree of compactness! How is such a packaging of DNA achieved without interrupting its functions? With a general viewpoint we may feel no hesitation in casually discussing on the ultimate truth, on God, on heaven and hell, but when with the same general viewpoint, and with the best of our capacity, we peep even into a single cell, we are bewildered seeing that even the simplest of life is so complex! It is life which incessantly teaches us what 'regulation' is! And see what a beauty of the sight, how promptly this simplest life happens to appear as if it is stretching thousands of its arms and saying to us "welcome into the world of pulsating pathways"!

Everything in a cell, in an organ, in an organism happens under a tight regulation, in a highly organized way due to which, all the cells and organs develop precisely and their specific roles come into play. It is only our conscious world where our thoughts often force us to believe that Nature is nothing but a chaos. Under the tight regulation, due to their properties, molecules behave in a certain way. DNA, being a double helix tends to further coil on itself; this is called 'supercoiling' because it is coiling of a coil. It is like coiling of a landline telephone cord that connects receiver to the dialler platform. This supercoiled DNA binds with some proteins and forms larger coils which are further coiled to form 'loops'. These loops bind with different other proteins and coil even further and further forming a highly dense and organized structure which we call as 'chromosome'.

Humans have two pairs of 23 different chromosomes thus a total of 46 chromosomes which are present in each 'somatic cell'. Somatic cells are not specialized for reproduction, cells known as 'gametes' serve this purpose. Gametes have only a single set of chromosomes, thus a total of 23 chromosomes. When two separate gametes, an 'egg' from mother and a 'sperm' from father fuse, the resulting 'zygote' receives a total of 46 chromosomes, 23 from each of the parents, how well contributed, organized and regulated an individual life is from its very beginning! Bacteria, on the other hand are single celled microorganisms, a bacterium has a single chromosome.

We have taken a glimpse of how DNA is organized into chromosomes. The concept of 'gene' arises when we look at DNA with a question "how does it function?" each DNA molecule, at first makes its copies, resulting into two daughter DNA molecules; the process is called as 'replication'. From DNA is synthesized RNA and the process is known as 'transcription'. RNA gives rise to Protein in a process called as 'translation'. These three processes construct 'the central dogma of life' or simply the central pathway for the flow of genetic information in living beings. Proteins are the products of DNA; they are the expressions of it and are the part of merely all the processes that take place in any cell. Genetic information, taking from the inheritance to the final expression of it, takes place in well defined units; we call these units 'genes'. Either call them the segments of DNA or say the parts of chromosome, genes are the units that encode for two functional macromolecules—RNA and Proteins.

When the sequence of genetic information in DNA is changed due to some environmental factors like radiations, chemicals or errors in DNA replication etc. we say that DNA is mutated. Mutation is an inheritable change in DNA which is often dangerous. Although each and every cell has a DNA repair system which quickly functions to efficiently repair the damaged parts of DNA, few changes may go unrepaired. A slight change is enough to cause 'cancer'—one of the deadliest and challenging diseases of present times. There are over hundred types of cancer which may arise due to infections, mutations, activation of cancer producing genes, poor health status and several other factors that ultimately affect the cell regulation.

Look at a patient who has come to know that his disease is incurable. Sit near him for a while, talk to him if possible and judge how depressed he has become. The severe effects produced by such diseases inevitably disturb the patient's mental balance and if the patient is not wise and humble by nature, the effects are more severe. Patients need care, people know it, yet there are millions of families who are unable to adapt to the change produced by diseases, and there are straight reasons for why it happens, the root of this problem in India seems to me to be—A significant number of families including patients are not well-acquainted with the medical facts regarding the disease and its progression. Life grows out of its contents, it rests upon its contents and even if the contents are certain, the arrangement of events is often uncertain. A balanced mental makeup is certainly needed by all of us, in every condition of life. We must strive to simplify the life if we can, be it in any phase, be it our own or of others, be it in one type of struggle or in different other types, be it a beginning or an end because we cannot live like a bacterium or like an organism that has no mental faculties; we are born and nurtured in a well-organized society.

6

I just kept on watching

Once again like always, a disciple pleaded with his master, "Master, you have a rich, powerful and magical life, but life is uncertain. Before you or I leave, please give me a mantra which on chanting gives a life like yours". The master replied, "My dear disciple, now the time has come that I must give you that mantra, it will certainly yield you a rich, powerful and magical life but beware of one thing, while chanting this mantra or going to chant this mantra, there must not flash any thought of monkey, if 'monkey' appears in your mind, the mantra will be of no use, I have no other mantra". Throughout my Ph.D. studies, I came across such masters and disciples working in different academic fields, but more frequently at different religious places where I wandered to find a person in which I could see my saintly Guru. But since I had no one at CIRG to whom I could confide my agony, I kept on watching silently to collect diverse facts being well-aware of the fact that it was an institute for research on goats and not an institute for research on gods.

In the above story, the master cunningly implants failure in to the greedy attempt of the disciple. He does it by emphasizing monkey with the mantra that he gives his disciple thus making the mantra and monkey two inseparable parts of a thought; thinking to avoid a thought of monkey itself will create an image of monkey. The irony is that the thought of monkey is antagonistic to the mantra; the mantra in the presence of the thought of monkey cannot have its effect. This picture of use of intelligence by the disciple and his master is important from several viewpoints including social, philosophical and scientific aspects.

Intelligence has several parts; however, the parts are not as distinct from one another as we see in solid objects. Let us look at two complementary parts among these parts: one part functions to set goals while the other part functions in pursuing them. It is awareness which spans both the parts and holds them together. The richness of goal and the efficacy with which it is being pursued depend on whether the goal and its pursuing was set with 'general awareness' or with 'special awareness'. The splitting of awareness into 'general' and 'special' denotes nothing other than the distinct degrees of awareness. When I look at the universe through the magnifying lenses of awareness, I find that any two things or objects in the universe may come into relation with each other by a fact—that they are the parts of a single fact.

It was April 2007. One day while going for lunch, in the way, I blurted out to my friend and roommate named Pravin who was walking with me "I have a feeling with a fuzzy image in my mind that a saintly man is about to appear before us within next few minutes". He looked at me and said, "May be, these days you are frequently visiting such men and meditating enough." We had hardly walked some 70 or 80 metres towards a turn when we saw a monk coming. "Oh my God!" We looked at each other in awe and wonder. Once again in my mind an image of a forthcoming event proved as a true foresight; this time to the maximum, the contents of mental image and a crude feeling followed by a blurred thought were uttered by me flawlessly! But once again I had no knowledge about the mechanism of all this; no explanation, no control studies.

I touched the monk's feet, dropped the idea of lunch and accompanied him to the river. In the way I came to know that he was a Buddhist monk. I asked him "What is the ultimate thing about life and this universe?", and this was the only question from the area of my philosophical curiosity that I asked to him, further I only listened to him, he spoke about many things. He said that the ultimate thing was not a matter of discussion. He told that he owned a monastery. Also, he told that he had been a class fellow of Atal Bihari Vajpayee, former Prime Minister of India. Atal Bihari Vajpayee (lovingly and respectfully known as Atal Ji) is one of the revered political leaders in Indian history. His idealistic life has been influential in the changing political circumstances.

I sat with the monk on the riverbank. He spoke about meditation and recited few lines. The language of these lines was not Sanskrit, though many words of these lines sounded like Sanskrit words. I was taught during my school days that Buddhist scriptures were written in many languages chiefly in Pāli; I asked "Is it Pāli? What is the source scripture of these lines?"

"Yes, the language of these lines is Pāli, I am speaking from *Pāli Tipitaka* (the Pāli canon) and from *Dhammapada*," he replied.

"Do you study the religious scriptures?" He asked.

"Yes, I am searching for answers to my questions," I replied.

"It's all right that you have questions but meditation is the most important thing," he emphasized.

"I love my questions, I cannot ignore them, they are my true companion. I don't think it is needed to repress the quest to know when one has realized that he has no thorough knowledge of Nature, of what Samadhi is, of what Yoga is, of what Kaivalya is," I said.

"The proper way to meditate is not to concentrate on the space between the eyebrows with closed eyes but to concentrate on the tip of the nose with open eyes," he said. While he was speaking, a 50 year old security man of the institute who knew the monk for a long time arrived there and said "Let me listen what knowledge is being imparted here".

"You fool! Go and do your work," reacted the monk.

"Yes, yes, I know well who the fool is!" the security man counteracted.

"If you know this, leave this place right now," the monk roared.

"Why? Do you own this place? I will not leave," reacting to the monk's remark he sat on the ground firmly. The monk stood up and began to walk towards the boat, I followed him silently. After walking around 50 metres, the monk said, "You are a good boy, beware of this type of people, they are too ignorant to understand the points, go child, go back," he said softly and boarded the boat.

"When and where can I meet you again?" I asked.

"Some day; I frequently travel through this route," he said and left.

I returned dejected. I had not expected that this would happen with the monk. Yet, my experiences had taught me not to make a hasty conclusion. A nascent conclusion needs a baby-like care from mighty observations. I thought that I must meet the monk again but I could not do more than waiting for it. I thought that before my next meeting with him, I must at least had a proper theoretical knowledge of Yoga and also at least some fundamental points of Buddhism so that I would understand him better or put my questions before him in a systematic way. After a few days of searching, I came to know that Indian philosophy had various schools chiefly Sāmkhya, Yoga, Nyaya, Vaisesika, Purva mimamsa, Vedanta, Jain, Buddhist and Cārvāka. Further I got to know that I should also study Sāmkhya in addition to Yoga as Yoga shares the viewpoint with Sāmkhya.

I purchased Sanskrit-Hindi versions of *Patanjali Yoga Darshan* (Patanjali's philosophy of Yoga; Gita Press Gorakhpur) commentary by Harikrishnadas Goyandaka and *Sāmkhyadarshanam* (Sāmkhya philosophy; Khemraj Shrikrishnadas press, Mumbai) commentary by Prabhudayalu. Since I had found the words 'Yoga' and 'Sāmkhya' frequently in *Bhagavad Gita*, I found it important to study *Bhagavad Gita* at the same time.

But I caught jaundice and went on one month's bed rest at my home. There I saw a book that my father had purchased—*Sthitaprajna Darshan* (Philosophy of steadfast wisdom) by Vinoba Bhave. As I opened it, it ran into my heart, I did nothing except reading this book again and again during the remaining 10 days of my rest. Influenced by Vinoba's life, his great philosophical simplification works, though not academic but to make the rare knowledge available for society, his movements like Bhoodan (Land-gift) movement, I determined that I would collect whatever else Vinoba had left for me to understand the essence of *Bhagavad Gita.* Within a week after returning at CIRG, I happened to visit Agra where I found another book by Vinoba—*Geeta Pravachane* (Talks on the Gita).

Dr. Ashok was happy with my performance in the laboratory. I had isolated over hundred bacteria belonging to different genera from the goats showing the symptoms of the diseases selected for my study. The goats belonged to the goat farms of CIRG and allied villages. The identification of each bacterium included microscopic evaluation and its characterization on the basis of biochemical and molecular methods. On the other hand, the collection of plant parts like leaves, bark, fruits, flowers, seeds etc. from the plants chosen for the study, then their drying, grinding followed by preparation of extracts by processing them in different organic solvents and water had been done. But the thing that made Dr. Ashok extremely happy was the way the things had happened while I was standardizing a technique known as Bioautography.

"The students that worked here before you had tried to standardize Thin Layer Chromatography-Bioautography (TLC-Bioautography). If you standardize it, the people will know about the calibre of my student, and I will assure them that they will see a brilliant thesis soon, how many days do you think it may take to have the results?" asked Dr. Ashok. "Around 15 days. Five days to go through the literature and the remaining, for experimental work." I said. "Ok then, get into it," he said. If we have a mixture of different compounds for example a mixture of different dyes, and we want to find out how many different dyes are present in this mixture or we want to separate them from one another in order to purify them, chromatography can be employed.

The word 'chromatography' is taken from Greek words *Chroma* (means 'colour') and *Graphein* (means 'to write'). Since the objects are colourful to our eyes, in chromatography we see the separated substances in terms of separation of colours. Chromatogram is the pattern of separated substances developed through chromatography. Bioautography is used to detect the antibacterial activity of components present in a mixture directly on a developed chromatogram. The substances to be tested in our study were plant extracts.

"So, what is the outcome?" Dr. Ashok asked on the 15th day.

"Sir, I have some tentative conclusions and speculations over the results that I have obtained," I said showing the papers, he examined the experimental set-up.

"What is now needed?" He asked.

"Sir, one week," I replied.

"OK, carry on," he said.

After one week he asked, "Is there any result?"

"Sir, the experiment is under process, tomorrow I will check it for the results," I replied.

Next day I informed him that the experiment had not shown the desired results. He examined one set of the run samples and said, "We have the results! It is done, what else are you waiting for?"

"Sir, as usual in this set-up are two things 'controls' and 'tests'. It is all right that the controls have shown the desired results, the tests were run in duplicate, also, it is fine that one set of the test samples has shown the results accordingly, but the problem is that the other set has not shown similar results, therefore the samples showing the desired results also must be taken for re-evaluation and confirmation, I need to repeat the test once again," I said.

"Oh! Yes but I am sure that in the next attempt we will have all the desired results, don't worry," he said. Subsequently we obtained the desired results.

One day in the afternoon, at the Institute's main entrance, I met the security man who had severe argument with the Buddhist monk few weeks ago. "I know you; your name is Deepak Dwivedi. I have heard your poems in *Kavi Sammelans*. The old man whose feet you touched that day belongs to a village across the river," he said.

"Great! Thanks for informing that the monk's village is nearby, I have been thinking about the probabilities of my next meeting with him, now I think that I would meet with him, you know my name too, you are well-aware of the persons around you," I said.

"He is not a great man, didn't you see how he behaved with me? I know him for many years, he belongs to scheduled caste," he said.

"The question is not whether or not someone is a great man, the question is do we possess the intellect that can identify the greatness? Does the wise hangs on only one class or caste? What caste do you belong?" I asked.

"I am from General category," he replied.

"Are all the people belonging to this class wise?" I asked.

"No," he replied.

"Then, do you think that the world is so squint that it cannot see a great in a simple man? The Nature is so incapable that it cannot produce a wise in a tribe? Why did this earth saw in the past, and in the present the elite and the affluent wandering in forests, on mountains, in search of right knowledge leaving their money, prestige, fun aside? What did they lack even if they belonged to high class families? If a society can divide itself to generate classes, it can also dissolve the classes to flow in love and peace," I said.

"Sir, you may think that a kid is trying to teach you, I am careful for my words should not pinch you. But for the happiness of a child like me, please clean up your thoughts once," I added and left.

But the widespread problems arising due to caste systems, racial discrimination, prejudiced views on land, language and economical classification began to pull my attention and gradually I became seriously concerned about the ways that could relieve the pains of world societies. Look at the diseased societies; can you prevent yourself from erupting at the sight of this distressing picture? And what a pity, these societies have greedy, pitiless and ambitious leaders who are putting their best efforts to avoid the cure! O heartless brains! O blind hearts! Have a walk into the world where your innocent progenies sing the song of values.

On blue planet, in the hues
Rising, moving to infuse
Come along, sing a song
Of eternal values

Growing learned to discern
Not to cling to concern
Eastern and Western
Northern and Southern
Above the transitive talent
And continental slues

Come along, sing a song
Of eternal values

Crowning them who know to bow
With no ego echo
Clapping for those who will go
Coming not long ago
On the land of languages
With the gems of virtues
Come along, sing a song
Of eternal values

Something as to give and take
For the gains for the sake
Let us reach the holy lake
Sliding eyelids at daybreak
And striving hard
Beating existential rues
Come along, sing a song
Of eternal values

Each of the Ph.D. students from CIRG had to face a serious problem at the University; no one in the University would pay any attention to the queries and problems of the students until the students knew that they were at a serious mistake; that they had forgotten to offer the concerned clerk, attendant and peon exactly that what they wanted first. And if, any of the students went to the higher authorities with a complaint of such an attitude and behaviour, he certainly realized that he had made a more serious mistake. And this was not with the CIRG students only, the students from other affiliated institutes and from University campus also had to deal with the same predicament. Only those who had their relatives or acquaintances in the campus, any strong references or strong positions would succeed in completing their courses without any hurdle. I along with my colleagues returned dismal from the University around sixty times. It took 20 months to have the report of University's research decision committee on our synopses. We almost forgot to talk about a 'proper channel'.

Dr. Ashok was worried taking my coming-on-front attitude; he believed that I was being targeted by the people again and again, and he wanted that I should not come on the front taking any issue, he was very serious about the successful completion of the work from which he had derived his high expectations, he therefore told me to keep quiet and to focus on work. Disagreeing with him and after a strong reaction to him on his viewpoint, I obeyed him for it was the best thing that I could do while feeling that perhaps I had hurt my guide.

As noticed worldwide, I also found an increased level of resistance to antibiotics in bacteria that I had isolated. An antibiotic is a substance produced by a microorganism which has the capacity to inhibit the growth of or to kill other microorganisms. The discovery of antibiotic Penicillin by Sir Alexander Fleming in 1928 was a milestone in modern medicine. Like many other scientific discoveries, this discovery was a portrait featuring 'serendipity welcomes watching brains'. Sir Fleming, by happenstance, noticed that a fungus *Penicillium notatum* had grown on a plate containing the bacterium *Staphylococcus* and that the bacterium was unable to grow in the area where the fungus had grown. After a series of experiments, he was able to show that the fungus possessed the substances that were able to destroy the bacterium. Sir Fleming along with Howard Florey and Ernst Boris Chain was awarded Nobel Prize in 1945. Antibiotics are now the mainstay of the treatment of infectious diseases.

But the treatment with antibiotics began to face with the problem of antibiotic resistance in bacteria. The problem is ever increasing as a result of which, many drugs that were previously effective are now ineffective. What antibiotics do is they interfere with the cellular pathways or structures in bacteria, for instance, targeting the cell wall (here the effect is bactericidal; killing the bacteria) or stop bacteria from multiplying by interfering with bacterial DNA replication, protein formation or other aspects of metabolism (here the effect is bacteriostatic). The resistance to a particular antibiotic develops by the process of evolution that we knew in the previous chapter (chapter 5)—each and every living organism struggles for its survival and the one, who under the effect of natural selection adapts to change, wins in the struggle. Bacteria evolve the ways to combat with the drugs used against them, as a result of which today we see Multi-drug resistant (MDR) and extensively drug resistant (XDR) bacteria, moreover, in case of one of the deadliest disease tuberculosis (TB), totally drug resistant (TDR) strains have recently been reported.

A major factor in the development of drug resistance has been the indiscriminate use of drugs. And we are responsible for it. This "we" includes both the patients and the physicians. I have not included the policymakers in this "we" because I have least expectations from them in this regard; I expect much from physicians and patients and I believe if they both use the antibiotics judiciously, we will have a rather practical solution to the most part of the problem. I don't know what the real situation in other countries is, but in India, the situation is complex and I believe that it has a strong resemblance with several other countries. Let us have a careful walk into the problem because the failure of treatment of infectious diseases is one the most severe problems of mankind.

Let us begin with the persons who dishonestly claim to have medical knowledge; they often succeed in luring the poor patients easily. It is pertinent to quote the 19th century philosopher Friedrich Wilhelm Nietzsche (15 October 1844-25 August 1900)—*"The most dangerous physicians are those born actors who imitate born physicians with a perfectly deceptive guile"*. They are unconcerned with the thorough diagnosis of the disease, the health status of the patient, thus they are ignorant of the fact that they are burdening their own environment with a greater risk.

The second thing is the cost and time involved in diagnosis of the disease, this is a checkpoint to a poor and illiterate but aware patient. But a crafty physician (another category of physicians apart from the born actors mentioned above) is able to cope with the issues of the cost and time involved in diagnosis in his own way. The physician knows that the patient cannot afford the cost, and if he is a private clinician, he is also aware of competition with other clinicians, and if the patient does not seem to have a serious history, he ignores the diagnosis part and makes a prescription, however, he prefers to prescribe the drugs that are of a high commission value to him. This crafty physician is not ignorant of the fact that he is burdening his environment with a greater risk, but he does not see this issue a matter of worry, he is quite of 'take it easy and forget type of mentality'.

The third thing is lack of attention to the problem of resistance. The seriousness of this factor is due to less aware attitude of educated and financially sound patients and failure in making an appropriate prescription by some of the real physicians. What is expected from such patients is they find a good physician, go through the diagnosis and complete the proper dosage regimen. The resistance develops due to incomplete as well as wrong treatment and due to reckless use of antibiotics where their use could be avoided. The physicians should try their best to avoid rashness and if not able to emphasize, they should at least make the patient aware of his specific regimen. I have no doubts that a true physician is more thorough and skilled than me. It is worth mentioning here what the 20th century physician John Albert Kolmer (24 April 1886-11 December 1962) had put on the dedication page of his book *Clinical Diagnosis by Laboratory examinations* published in 1944—*"No clinical laboratory examination can be better than the thoroughness and skill with which it is conducted"*.

For few weeks in the hostel, mysterious activities had become a matter of daily discussion among boys. Merely each of us in the hostel and many other boys and girls in different laboratories had been the victims of theft. The reality showed us what none of us had expected, nor had ever seen before; a handsome and muscular M. Sc. student, who was staying in the hostel, was the culprit and was caught red-handed. And the way he was caught by us involved the help from a *tantric* that lived one km away from the institute, in the same village. The word tantric refers to both a body of system described in scriptures and the one who practices Tantra. The word *tantra* is used to refer a set of doctrines and ritualistic practices for obtaining spiritual enlightenment. Along with many of the world religions that possess similar practices with different names and set of principles, tantric doctrines are a part of Hinduism and Buddhism.

Some of the villagers had advised the Ph.D. students to go to the tantric to find out the thief, the idea sounded strange but intriguing; consequently all the boys from hostel except me and another Ph.D. student went to the tantric; I was meditating at the time the boys were leaving, therefore they did not disturb me. Incidentally, the thief himself took a keen interest in finding out the thief; therefore he was among the boys who went to the tantric. The tantric said "Two of you have not come to me" and within minutes he exclaimed pointing out at the thief among the students "You are a thief, you have committed theft!"

All the boys including the thief returned astonished. They narrated the story to me; I decided to go to the tantric coming Sunday. Meanwhile, after a plan set-up, the theft was proven as the culprit was caught while attempting to burgle once again. Then he disappeared from the hostel and was fortunate enough as his tenure was ended. On Sunday, I went to the tantric with some flowers to offer him; after a brief introduction, he seated me on the cot he was sitting on. I asked, "What is the ultimate thing about this world, about life?" he replied in a different way; he began to recite a long mantra loudly beating his chest.

"Why these two opposite activities?" I asked. He repeated the same thing.

"OK, but why these two opposite activities?" I insisted, he again repeated the same thing.

"OK, should I take your leave now?" I stood up; he said winking one eye "Oh! No, sit, you are getting angry, have some tea," and shouted facing the kitchen of his house "Prepare two cups of tea, a special guest has come".

"Did you understand what I did in response to your question?" He asked.

"Yes, you invoked a deity reciting the mantra and tried its vigour on me," I replied.

"Oh! This means you know the things well," he said winking his eyes.

"But I don't understand why these two opposite activities?" I questioned once again.

"Don't try to understand, I also don't try to understand," he said and closed the chapter.

"I had come here with high expectations," I confided. He extended the tea and had a sip.

"I have nothing to do with the Vedas or Sanskrit etc," he said.

"But the language of the mantra that you recited was Sanskrit, it was devotional but you recited it in an aggressive way, that's why I asked you again and again why these two opposite activities? do you understand the meaning of this mantra?" I dared to ask.

"Oh! It is a tough day today," he expressed holding his head into his palm, and then began to giggle.

"OK, no problem, but many of your sudden remarks often prove true and exact, the people say this, and I have also witnessed the same thing last week. Can I know what is behind this quality of yours?" I asked.

"I don't know, I just speak and it goes true," he replied.

"OK. When did you first knew about this quality?" I asked.

"I had died a few years ago. And in my previous life, I had no such quality, few hours later after my death, I got a second life with this quality," he replied.

"My God! I had heard and read about these things but I had not imagined that I would meet such a person. What happened during the hours between your death and regaining the life?" I asked.

"I remember people gathered around my dead body, rest I don't remember," he replied.

"OK, have you tried to find out the purpose of this second life?" I asked.

"No," he replied.

"Do you think that it is important to know the purpose of your life?" I asked.

"No," he replied.

"OK, should I take your leave now?" I asked.

"Yes," he said. I greeted him and left. I went to him twice again and noticed his activities; subsequently I dropped the idea of meeting him again.

Further, in a period of around two months, I visited few villages where I met another 5 people said to possess mysterious powers, each of them possessed a different way of practice, two of them were tantric. Each of them tried to make the things more and more mysterious, one of them even prepared to launch an assault on me, whatever I went through in this direction, I was able to register two points in the book of my belief which I take very seriously: first, Psychic powers are real powers, they are systematic, principle based and explainable. Second, among those who possess such powers, a few have a systematic and universally applicable knowledge of them.

At the end of 2007, I shifted at Agra, 35 km away from CIRG. The reason behind shifting was my health. I had grown ultra-thin due to lack of proper arrangement of food and repeated infections. The condition was worsened due to unconditional physical and mental exercising. There was no problem in commuting as the institute provided excellent bus facility. It never took more than two hours a day in travelling. These two hours of the day were the golden hours for me, and in the beautiful, blue coloured bus driven by a disciplined, around 50 year old driver named Amar Singh, I could comfortably know about great men by going through several books in next two years. Many thanks to Amar Singh ji.

I am also thankful to my colleagues who travelled with me always being aware of my seat (the last, left window seat) and showed a great respect to my admiration for the past greats. Now, it is of great practical importance to confide that during every fresh session of dissertation training periods, M.Sc. students mostly girls, for few initial weeks of their joining would murmur "Deepak sir is wasting his golden period, he should enjoy these days, books are not everything" then, in the middle of their training period, they would murmur "Why don't such thoughts arise in our brains? Perhaps we should follow him" and finally towards the end of their training, few of them would request me "Sir, you have many books, can we take one of them home to read?" I don't know whether they read those books or not but they never returned them.

I am basically a fool; however, baffled by the modern fashion, sometimes I tend to behave like an intelligent person; I feel embarrassed after ignoring the facts. I learn things with a special awareness to my follies; take much time in observing to find out whether I have observed properly or not thinking that it is the quickest way to meet a sustainable conclusion; prefer walking to reach a distant location thinking that the paths love teaching those who stay longer with them; I walk with a pace often slower than that of my childhood as I think that if I survive, I would be able to achieve the fastest of my pace in my later life.

Nothing is more complicating to a book than that it is left half-read. And as a matter of fact, most of the books are complicated this way. The knowledge acquired through this approach results in a paralyzed body of conclusion. If you find a book to be dull and if you feel that you are unable to understand what it is trying to convey, try to find out the point of view that you are missing, just go through the way it speaks, it will clasp you, it will not leave you alone and you will not leave it aside. Books follow the rules, the rules are not their limitations; they are their ornaments with which they adorn themselves, for the joy of a reader, for the shine of a learning eye.

On the other hand, there is something for a careful consideration of authors; they must avoid producing careless and prejudiced literature and must follow a standard way of expression. It is better to delay binding the book for the sake of lucidity because it is rashness that debilitates the efforts of author, there are already established terms for such a rash mental somersault; one of them is 'smattering' which means superficial knowledge. Every juvenile knows a fact that it is better having a thorough knowledge of a single point than having a superficial knowledge of several lines. Yet, as a matter of worry, the modern history is full of such attempts; to a greater irony, it continues to happen in the present era of specialized studies. Imagine of just two consequences on two individual readers who with little money in their hands to get a proper understanding of facts, had any how managed to purchase books, one for each; one of them has learnt not to waste money and time further while the other one is happy with reading a blindly assembled interpretation of real things!

In addition to the problem discussed above, when I came to know about the doctrines belonging to different schools of Indian philosophy especially Sāmkhya, Yoga and Vedanta, I learnt few important facts regarding the source and type of literature; I knew about what is called primary, secondary and tertiary source of information or literature. While going through different commentaries on these texts, I found that they varied at several points depending on the emphases of commentators. The more I ran into the problem, the more I got confused but the longer I stayed with a particular school of thought, the more it mesmerized me.

The definition of primary, secondary and tertiary source of information or literature may vary depending upon the subject of enquiry. In a broad sense, original information or works are primary sources; they are usually the first appearances of facts or documents. Commentaries, critics, reviews, meta-analyses and text books are secondary sources as they compile, evaluate or interpret the information presented by primary sources. Encyclopaedias, bibliographies, reference books and directories are tertiary sources (however they may be in some conditions considered as secondary sources). For instance, this book is a combination of primary, secondary and tertiary information. Since it is an autobiography with expressions of my individual thoughts, it is a primary source, and since it gives an analytical view on issues and points previously discussed by several people in past, it is a secondary source and because it cites and discusses about the primary and secondary sources, it is in part a tertiary source of information.

I fell in great admiration for deep realizations of ancient intellectuals which never took the ambitious form of pursuits that seems to be prevalent among present day brains. I began to think "If the purpose with which our ancestors lived was so sacred so great, is it not true that the modern world is struggling with numerous unrealized modes of living?" If asked what is to be done today with religion? I would promptly say "Efforts to preserve and understand the oldest forms of world religion". I would say this because I believe that the proper knowledge of these views to coming generations is as important as the knowledge about ancient cave arts and fossil records to understand our ancestors better, to understand the serious practical realizations of the times when the environment was not polluted.

The intellectual wealth of Ancient India is now admired by most of the world thinkers. In addition to the efforts of Indian scholars of modern age, the serious and admirable efforts of the scholars from different areas of world illuminated this great wealth. The literary critics have been central to these efforts. With a side effect of producing a bitter feeling, the critique favours the truth. Though unavoidable to a man with general viewpoint, such feeling is subsidiary to a man with special viewpoint because to him his ego is not as important as it is important to establish the facts, the truth for he knows the essence of any systematic unprejudiced enquiry.

While striving to find answers, it is important to welcome newer questions whether arose in an individual's mind or brought into focus by another individual. Similar questions, similar point of views arising independently among different people show the universality of human reasoning and experiences. Different ways of studies like science, philosophy, arts etc have immensely contributed to establish universal facts. When an individual finds that he has something universal in him, he must strive to stay watchful as long as possible on the boundary between his individuality and universality because in doing so he will not feel dizzy while looking at the 'dichotomy' of his existence and the 'monotonous fixedness' of his experience.

A traveller and a village
Together
Rapture
Closer
Slur
Opening village, entering traveller

Surroundings
Bloom
Air
Surging
Vital
Picturesque village, waking traveller

The traveller
The village
Here and there
This and that
And this and that
Saying village, listening traveller

Things
Ways
Position

Boundaries
Moment
Cuddling village, passing traveller

Creeping silence
Gazing at
The village
Lost in traveller
The traveller
Beside the village

When we come to know about universal facts and principles connecting them, we are fascinated by scientific explanations. In the modern era of specialized studies science, arts, religion and philosophy are separate systems. With religious belief being a distinct aspect due to its social nature, the rest three systems in their fundamental forms remain inseparable in a man; anyone whether a man with general viewpoint or a man with special viewpoint; anyone whether religious or not; anyone whether he is aware of the fact or not naturally is an artist, a scientist and a philosopher. The difference is in the degree of development of these systems in him, the extent of expression of these systems in him depending on the environment. Those who turn more scientific don't lose the elements of arts and philosophy in them, they always have them; but the degree of expression of arts and philosophy in them is dominated by that of scientific methodology. The same thing applies for all the remaining systems. Whatever the system, 'systematization' is the central principle of each of them.

One of the greatest uses of human intellect culminated in ethics. I don't think anyone would disagree with this assertion; even if he who is careless enough of his behaviour to others expects a gentle behaviour to him from them, on the other hand those who prefer loving dog to man often hate to be loved as a dog. Ethics is also known as moral philosophy, and it is ethics which gives philosophy a unique colour and versatility of its implications. It is not strange that the precise understanding of ethics is generally found among philosophers only. In a general viewpoint, ethics is not taken seriously, and if taken, is often confused with religious imitations and partially psychological healing tactics of greedy people. Another thing to this direction is that a significant proportion of human population can only understand the language of law. It is of great practical importance to look at what one of the most influential philosophers of all time Immanuel Kant (22 April 1724-12 February 1804) had said *"In law a man is guilty when he violates rights of others, in ethics he is guilty if he only thinks of doing so"*.

In April 2008, I happened to meet the Buddhist monk again, this time without any prior intuition or thought. I was going to laboratory when I saw him going towards the river. I ran to him and requested to visit the hostel thinking that it would prove as a memorable occasion for us to learn something directly from him. My thinking was the result of my realization of previous puzzles created by varying views of different authors on the same points. It had required repeated attempts to go through these views in order to understand and harmonize the conclusions and I could manage to find time for this task with a lot of difficulties. He accepted my request and spent half an hour in the hostel. None of the students, due to his individual reasons, could stay with the monk while I remained busy in arranging snack for him as he was looking tired. No discussions took place, no further meeting occurred, the only satisfaction to me was that I was able to serve him food.

I received no fellowships during my Ph.D. I always received money from my home until I got a research fellowship few months after the submission of my Ph.D. thesis in August 2010. But in May 2008, I was getting worried for one reason and that was I estimated that my work was going to be completed in next two years or so and the thesis binding, submission, panel selection, viva voce and finally the award of degree would need a lot of money, already I was having a feeling of putting much financial burden on my family, I decided to go for home tuitions. By the way, a scientist working at CIRG who lived at Agra, offered me to teach his son one hour a day for which he finally agreed to pay Rs 1100/—per month. Few months later a younger colleague of mine named Manish Dubey, then a M.Sc. student, took me to teach a class 9th girl student who lived in a big multi-storey building; fortunately the family found me appropriate for their purpose. Manish possessed a sound knowledge of Yogic exercises; he served as a Yoga tutor to the family while I, as a science tutor to the girl, each of us received Rs 3000/—per month.

I was motivated by my student, however, she was unaware of both the things; that I had derived motivation from her and that she possessed something that could motivate someone. Though, I was having a busy day schedule that did not allow me to find more than 30 minutes of rest between 8AM to 10 PM, my student seemed busier than me. Besides school hours, she had separate teachers for different other subjects including dance and music. The important thing was she was extraordinarily composed. Her mental balance was a lesson for me and imitating this balance, I could stay longer with the fundamental principles discussed in class 9th text books. Physics, Chemistry, Mathematics, Biology, and so and so began to appear to me as the colourful inks of a fact book of Nature. Galileo, Lavoisier, Pythagoras, Aristotle along with hundreds of greats seemed to me strolling with one another in a grand circular path. Oh, what a sight! What a wonderful sight! The days of celebration, subjects of studies dancing around the students of Nature!

Great microbiologist, Louis Pasteur (27 December 1822-28 September 1895) had said *"Fortune favours the prepared mind"*. You might have noticed or if not, you may notice it now that when I said earlier in this chapter 'serendipity welcomes watching brains' was constructed upon this aphorism. The word 'serendipity' relates to 'fortune', 'welcomes' relates to 'favours' and 'watching brains' relates to 'prepared mind'. It was a delineation of the mechanism I saw looking at the Pasteur's belief expressed in this particular aphorism. At the end of 2008, I felt a favourable gust of fortune around me: While running through the great works of Sir Isaac Newton (25 December 1642-20 March 1727) and the immense contributions of the scientists that worked in classical and quantum mechanics, I got to read *A brief history of time—From the big bang to black holes* (1988) written by physicist Stephen Hawking. The book is one of my favourites. Hawking is an example of how a challenging life can produce a rare simplification. Am I not fortunate enough to boast that I live in the age in which lives Stephen Hawking?

Perhaps the most genius, the most influential, the timeliest recognized scientist that we ever had was Albert Einstein (14 March 1879-18 April 1955). He is the hero of most of the intellectuals. He is the hero of a fool like me too. You should not underestimate my assertion; I have reasons for how a fool can rightfully call him as his hero. He made me believe on the power of imagination, he made me to aim at originality and he made me to stay longer with concepts. I never found any of his remarks to be rash. However, one of his remarks that initially puzzled me was "If the facts don't fit the theory, change the facts" because we know that facts are facts they cannot be changed. If one knows that Einstein is a genius, and that the issue is serious, greats never make fun with the things that may create serious confusions there is no possibility that the remark is a fun. Then what is meant by saying "change the facts"? Einstein was a great teacher who defined the power of 'theory', the rational imagination. To him, theory was very important, and in that light he said don't discard the theory; apply it to another set of facts that may fit to it. If this was what he exactly meant, 'change the facts' meant 'replace the facts'.

During the favourable gust of fortune, I got to read *Autobiography of a Yogi*, an autobiography of Paramahansa Yogananda (January 5, 1893-March 7, 1952). Yogananda's life clasped me such that I took two day's leave and postponed everything to live with Yogananda. I met with a miraculous and disciplined life that endeavoured to explain the underlying principles. Later I came to know that Yogananda's life had proved to be life changing to several influential persons. His life overwhelmed the long yearning of mine, and I asked in loneliness again and again to myself, to the blowing wind, to the glaring sun, to the glowing moon "Is there any Yukteswar Giri or Lahiri Mahasaya for me?" (Yukteswar Giri was Yogananda's Guru, and Lahiri Mahasaya was Yukteswar's).

It was 2009. I had no personal computer until 2010, and due to my tight schedule, it was not possible for me to go cybercafé other than Sunday or any holiday. This whole year, I could not walk much in the direction of finding answers to my loving questions. I got busy in presenting and publishing few results of my Ph.D. work. Moreover I paid much attention towards postgraduate students in the laboratory because I had witnessed a painful time during my postgraduate research and did not want in any case to see the innocent beginners suffering for anything. I saw a great improvement in the functioning of each and every student and it gave me a lot of satisfaction. My Pains had taught me well how to protect someone from them.

I had been practicing astrology, though occasionally, the only aim was just to keep alive a knowledge that was gifted to me by the great man. Few people began to advise me that I should open an office for astrological consultancy which would be a high profit business; one of such interesting advices was "Why to die for a job that will give you little money for a daylong exercise? Truly speaking if you say today that an office is to be open, I promise tomorrow you will sit there and will see a long line of people outside it". My only answer to such advices was "There are two clear reasons for why it is not possible: first; it is not possible for me to accept money or any benefit in return and second; I have no thorough knowledge of this science, I can make just interpretations, I know no calculations, and I don't know the real mechanisms, though I am following a set of rules yet it is a matter of chance that my predictions are working".

Whether the above was a bold assertion or not, the following is certainly bold: *This is how for many of us several types of knowledge are not only useless but also a serious blunder of mankind; Give birth to a child, nurture it poor, and murder it fearing for an infectious disease it may spread, thereafter invent a ploy to prove that the birth was illegal, a terrible accident! For more than a year I struggled to understand Avidyā while studying Advaita. When an atheist ridicules at a theist, a psychologist at a metaphysician, a scientist at a philosopher, a politician at a social activist, Avidyā becomes evident. Blind and squint attacks are never a substitute to dignified use of logic. On the grounds of practicality, the human brain always welcomes the authentic ways; are you not able to see a possibility of peaceful stay of a theist into a philosopher, a magician into a scientist?*

Astrology is not a figment of imagination. The only problem with it is that it is often contaminated by the mysterious elements. There is a need to employ statistical analyses, quantitative as well as qualitative research in to modern astrology. I am aware of the court orders that different individual organizations in India have been issued in past few years. I request to authorities that they run a project on astrology aiming at the validation of Indigenous Intellectual Knowledge. I believe that India has a great wealth of brilliant scholars that are unknown to masses and hope that a search for such scholars along with a systematic study on the subject will provide meaningful insights for whether astrology can be accepted to be incorporated into education system or not.

In April 2009, while searching for a volume of *The Bergey's Manual of Determinative Bacteriology* in the Institute's library, I saw *The Encyclopaedia of world religion.* After finishing my work, before leaving the library, I took the encyclopaedia out of shelf and just opened it from the middle of it; I saw—*"Take up one idea. Make that one idea your life—think of it, dream of it, live on that idea. Let the brain, muscles, nerves, every part of your body be full of that idea, and just leave every other idea alone. This is the way to success"*—Swami Vivekananda.

True, a moment is enough to change a life, a moment that brings you the power of sun, a moment that takes you away from the pushes of your past, a moment when a purpose shines in you. (On the moment when I saw Vivekananda's message).

Rarely a life reaches to you with such a velocity, rarely a life holds you so firmly, rarely a life works with such a responsibility (On Vivekananda's life and his works).

Great is the purpose that awakes a sleeping one, great is he who sleeps with it to rise with it again and again (On Vivekananda's purpose).

What an occasion, a fool clearly sees that he is more than a fool! (On Vivekananda's influence on me).

Towards the end of 2009, Dr. Ashok came to know about my activities and views; puzzled and little worried, he asked, "What is the purpose behind this vigorous exercising? Do you think it is meaningful? Are you focused? How do you balance all this?"

"Sir, Purpose is getting clear slowly, a yearning to know is clear, it is meaningful, I think I am focused, I don't try to balance the things much, I just do what is most urgent, the things are automatically getting balanced," I replied.

"That is all right but what about your career? What have you planned?" He asked.

"Sir, nothing in this direction, but enough in the direction of my yearning, I have to work very hard and carefully," I replied.

"I am unable to understand," he said.

"Sir, I am also trying to understand the things," I replied.

"Are you in the mood to submit your Ph.D. thesis?" He asked.

"Yes Sir, but I don't want to compile it rashly, next few months are crucial, I am well set to put the best of my efforts," I replied.

"Do it in time, I know you are already a scientist but don't get so absorbed that you exceed the time-limits," he said.

"Yes sir, just two or three months and then I will start writing," I said.

"OK, but fast," he said.

Dr. Ashok examined much in me, he peeped into every aspect of my life, silently, and consequently, he always overestimated me. In the present research atmosphere in India, Dr. Ashok, as a guide, is an exception. Very few, very few of research guides in India admire their students, and very few dare to metamorphose their students in to visionaries. There is no dearth of scientists in India that carry out top class research and bring out extraordinary results, though they are great achievers, and they have numerous publications in high impact factor journals, numerous awards, and a long list of patents, many of them have no time to know about the vision of the founders of the organizations with which they work. Students are not cheap labour; they are the only hope of future. I was brilliantly lucky that Dr. Ashok accepted me as his student, and I saw how much elated my guide felt at the moment I left the Institute. I submitted the thesis in August 2010, however in haste. Dr. Ashok was the happiest man; I had completed my work, we were able to file few patents with the antibacterial herbal preparations and could commercialize few products.

As happened with my colleagues, it took a variety of my efforts to get my viva voce conducted by the University. I applied several types of systematic knowledge like guessing the mood of the concerned clerk, tracing the working chain of clerk-attendant-peon and their daily tuning, finding out the reason why the concerned chain behaved in an unexpected way, asking the same question regarding the status of the process in varied ways to find out the real status of the things etc. in this enterprise. The efforts worked and I got my viva voce conducted in the 9th month from the month of thesis submission; in May 2011.

Meanwhile, few important things happened in my life, especially January onwards. In January, I fell dangerously ill; I was given a wrong treatment of fever that aggravated my condition. I did not inform my family. On day 4th during that treatment, around 8 PM I woke up with a thirst but could not rise from the bed, I tried as hard as possible but failed to make myself sit. I realized that it was not easy to leave bed that night, ok, I planned to rise. I looked at clock, it was 8:50, I decided that I would make an effort at its maximum at 9:00. I felt a *conditioning of will*, stronger and stronger then at 9:00 I realised that not such a big effort was needed, it was easy to leave the bed, and yes I rose from bed almost effortlessly.

But as I walked towards the water jug which was on the table, I fainted. I regained consciousness after an hour or so; "Oh! The ground is cold; I was on the ground for an hour!" I thought and tried to walk but fainted again, and the same thing happened once again but this time I could manage to walk four feet to the door of my room before fainting, this banging caught the attention of my landlord's family. They made me sit and I regained consciousness with a blurred vision, I said "Vomiting" they took me to bathroom where I vomited, then gave me water and took me to my bed; I said I was fine and requested them not to phone my home. Lying on bed I began to think "Am I going to die?" "What if I die?" "My family will not be able to bear it" "I have done nothing for my family" and a fear governed by the thoughts of attachment to family overtook me. Subsequently I thought "Where had this fear been lying hidden in me? Am I not the same boy who never thought like this while stepping into the waist-high river, while wandering in the forests and desolate places in search of my Guru?"

I whimpered. "What is the outcome of this life? Is my life worth? It always puts its best to keep itself alive, for what? Is it entrapped between *Karma* and *Moksha* or *Nirvana* or *Kaivalya*?" I kept thinking until I fell asleep. Next morning I moved to CIRG hostel where I stayed for next 15 days under the care of the boys. After I recovered I decided to observe the things that I had been through in order to know as early as possible what my life has been worth living for. After a watchful day I found that this question itself needs a critical analysis. Not as a skeptic but as a scientist I sought to collect the facts that would guide me to establish a clear line between the events of 'my' life and the events of life.

The history is the best way to future. I thought that it was most appropriate to look at my past and to look at history all around but I was also aware of the fact that it was not possible for an individual life to walk all along the vast land of history. After 5 days of serious thinking, I became sure that my life had been worth for almost nothing and realized that my conscious life had seen several greats. I decided to stay with my past up until the present and at the same time to walk into the lives of past greats who had contributed in the concerned areas of my curiosities in order to learn from them so that I would be able to make my life worth for something. Since I was already having a big tree of questions in me which was still growing, I walked into the direction guided by these questions. Now it was all about life and lives.

I love seeing birds flying especially flights of young ones though I don't like to touch them. One day while sitting on the roof of my house, I saw what I started this book with; a dove which I knew very well, in the hands of few people. I wrote down about this event and thought of systematizing the accounts and my views on paper, however as yet, I did not dare to think that I deserve writing a book at all.

In February, except my family and my friends whom you met in chapter 3 and 4, almost everyone who came to know that I was busy in reading neuroscience, fundamentals of physics and chemistry and western philosophy began to criticise me, some of them criticised me ferociously, never forgetting to ridicule me on each and every meeting. I asked to each of them "Am I doing anything wrong?" their egoistic reactions hurt me, and for next few months it discouraged me much to see that my yearning to find answers to my loving questions received no one's appreciation. Naturally, I grew more and more concerned with what I was moving towards. In almost every individual's life come times when he finds that no one is able to understand him at the moment, still, during such miseries, few are able to precisely hope that at some another moment the world will feel glad to understand them.

In March, I appeared in an interview for the post of Senior Research Fellow (SRF) to work in Outreach program on zoonotic diseases—Verocytotoxic *E.coli* at Pt. Deen Dayal Upadhyay Pashu Chikitsa Vigyan Vishwavidyalaya Evam Go Anusandhan Sansthan (DUVASU), Mathura. It was my first interview for a job. Few days after the interview, I began to receive threatening phone calls that if I joined the post, I would be thrashed so thoroughly that I would not be able to continue the job. Thus before receiving any official letter, I came to know that I was selected. The only thing that I had learnt from the dangers of life was 'to watch'. I joined the post and worked there until I got selected as a Post Doctoral Fellow (PDF) in Indian Council of Medical Research (ICMR).

In July 2011, I joined National JALMA Institute for Leprosy and Other Mycobacterial Diseasesas (NJILOMD), Agra (ICMR) as PDF to work on leprosy. Ajay, one of my Ph.D. colleagues, who was already working as a PDF, had made me aware of the fellowship program and helped me at crucial steps of the procedure.

I started writing this book in September 2011 and completed first chapter with a lot of difficulty as I was given a bunch of responsibilities at work that required at least 10 hours a day functioning and because my English knowledge was at least poor if not very poor, yet I chose English because of the subject of the book; it was easier to me to transliterate the Hindi expressions into English but very difficult to transliterate the English scientific and philosophical terminology into Hindi, another reason was that I was narrating to the world rather than a nation. Thread had been special to me since my adolescence and also my companion at almost every turn of my life, moreover I was aware that the word 'thread' had served as a versatile metaphor for numerous expressions since ancient times, therefore I saw life through thread.

Acronym 'JALMA' in National JALMA Institute for Leprosy and Other Mycobacterial Diseasesas (NJILOMD) is special to India. JALMA stands for *Japan Leprosy Mission for Asia,* a Japanese voluntary organisation. NJILOMD was originally established as *The India Centre of JALMA* in 1964 and in 1976, the centre was given to Indian government.

Leprosy is one of the oldest diseases of mankind; the disease was known in ancient China, Egypt and India. This disease has been a disease of serious misunderstandings as a result of which the patients suffered in most painful ways. In all the civilizations, in all the nations leprosy has been a mark of disgrace, a curse, or a punishment from deities. Leprosy patient was not allowed to join the social gatherings, to use the common roads, to touch the items used by healthy people, to the societies he was an ill-omened, luckless man. In 1873, Dr. Armauer Hansen (July 1841-February 1912) of Norway discovered that leprosy was caused by *Mycobacterium leprae*, a bacterium; this was a revolutionary discovery that paved the way for the treatment of this disease.

'JALMA' is special to India because it is an example of good faith; service from the doctors of Japan to the patients of India. It was an initiative in India. In 1959, Matsuki Miyazaki from Japan came to India to study the leprosy situation here. In 1964, he established the India centre of JALMA, and assumed the post of the first director. After his death Mitsugu Nishiura served as the second director of the centre. The dedication of the teams led by these two greats not only served to treat the patients in the times when the social stigma was prevalent in societies but also taught the societies to understand the despondency of a diseased man. A student from India salutes you Nippon!

'Stress' is special to me, in all of its possible meanings. Life is a rare example of stress. Modern man has seen it very closely. It is March of 2013 and I am walking through a rather challenging phase of my life; I am trying hard to find the traces of simplicity in selfish lobbied people who are clever enough to prove that they have served humanity; I am trying hard because I see the students including me frequently hurt by ill motives and unbearable reactions of these powerful people who have been careful enough to gather strong evidences all along their lives to prove that they have served loyally; I am trying hard because I could discover few honest people and because there are many who possess vulnerable innocence. Mountainous goal, growing challenges, physical, emotional. Fortunate are those who breathe a thought "stay firm to watch".

7

Something

In this complex world, I travel through an aisle between two innocent thoughts—that every goal is achievable and that in the end, I own nothing.

Swathed in meaning cascades
Embodied life
Splashing experiences
Developed end
In the heart of quiet space
Wanders restless earth

Man possesses viewpoint which may be 'general' or 'special'. This chapter is a dialogue between these two viewpoints; between these two tendencies of a man. You might think "why is a dialogue between the viewpoints needed?" I hope you are able to find a tentative answer at the moment, but I think it would be unprecedentedly enthralling to relook at this innocent question when this chapter comes to a full stop. Man has something with which he strives to explore everything and thus *something* becomes more important than what he calls as *everything*.

The knowledge of this *something* concerns the brain and thus the mind. I have put the two words 'brain' and 'mind' together; in a general viewpoint it appears that the two are same and could be used interchangeably, however in a special viewpoint they are not substitutes for each other neither they denote precisely the same thing, moreover, if you are a modern scientist with a physiological orientation you may prefer to use 'brain' and if a classical philosopher, you may prefer to use 'mind'. I have put the two together because I, like many modern neuroscientists, psychologists, cognitive scientists and many others believe that both, the brain and the mind are specific frameworks for mental phenomena and that the brain is the basis of the mind.

Whatever this chapter has grown upon is not new except the sequence of points. The points are derived from already established facts and theories; the sources of these points are cited according to the context, however, as often happens in systematization, it is possible that certain assumptions that I make during the discussion, or my views, seem to be just a slightly different expression or even a replica of few points that are already established by someone else that I am not aware of, I would like to proclaim that this book does not have any interest towards more than minimum necessary ownership, the journey of this book has a deep influence on me, and I have felt in its all vigour, every bit of this influence day and night for more than two years.

As a part of this chapter which is as analytical as narrative, I have tried to develop my views on few events of my life that I take seriously. Respecting the standard way of theorizing, I restrict myself to take my views as *systematic assumptions* and put this book before the present thinkers who, I believe, will certainly value it more than I could. Now it is appropriate to take pleasure in saying that I respect my foolishness for it has taught me two things; to wonder shyly and to imagine boldly.

My life is more interesting than me. This does not mean that *I* and *my life* are two separate things. But when I think this way for the sake of an analysis, I find that my life and my *'I'ness* both are two interdependent elements of something which is localized and that my life is more fundamental than my *'I'ness*. I take this something as *a series of concurrent events*. Then, to analyze this view, I move deeper and find that the source of *'I'ness* is the body and more specifically the brain. I move further and peep into the brain, I find that the *'I'ness* is an expression of self-awareness.

But with a general viewpoint, I see uncountable life forms around me that possess no brains, no awareness, and no self-awareness yet they seem to be very much interested to survive as long as possible. My body is made of trillions of cells, they are also brainless, possess no awareness, and no self-awareness, yet they function in a certain way, in a certain direction, how? Why? What is behind it, the ground of it, the cause, the cause of life and life events? The answers range from 'God' to 'Nature'. 'Consciousness' spans these ranges and is perhaps the most preferred term for the fundamental principle of mental phenomena, the fundamental principle of life and to many of us, the supreme reality.

Before we set out to meet particular answers, we need to look at the available ways to proceed through. There are two popular ways of systematic enquiry: top-down and bottom-up. Top-down way explores the details, or in other words, it starts with a higher or complex order of the things and progresses towards the lower or simpler orders to generate details at fundamental levels, it is deductive reasoning and thus analytical. Bottom-up way takes the opposite progression; it starts with the details of elements and progresses towards the higher orders. It is inductive reasoning and thus, in a sense, takes the way of synthesis. Considering the nature of the above questions, we will begin with the top-down way and then while making certain systematic assumptions, we will progress bottom-up.

Since we have walked together along a thread, into a life which I call mine, we are familiar with the questions that my brain touched. We have seen chapter by chapter how these questions arose and proved their importance. What these questions may finally give us is still uncertain, but their occurrence in a human brain suggests me three things: First, questions explore and establish a sequence of events and simplify the things to make them comprehensible to brain, they guide and they serve. Second, human brain not only simplifies, solves and regulates the things but also creates and enjoys puzzles. Third, the brain, in a sense appears to be an intelligent seer and performer, in another sense it appears to be the product and part of Nature, and in yet another sense it is a natural mirror of Nature; the mirror through which Nature sees itself.

What is Nature? The question is not an easy one, but I, like many of us, believe that it is much easier to define Nature than defining God. Yet, something becomes very important regarding our understanding of Nature;

Deepak Dwivedi

I hope that almost each and every individual will agree to a fact that we, due to the limitations of our intellect or our individual nature are apt to underestimate the grand Nature, and often act rashly in making the most important conclusions. In the world full of proper ways, the most serious mistakes are made by the intellectual beings. The irony is that our development teaches us that the meanings must grow and so we are busy in digging a vast trench separating the meaningful from the meaningless.

Historically and naturally the Nature has been defined in many ways, most importantly in three ways: Religiously, philosophically and scientifically. The religious views vary the most on the question of Nature yet most of them agree on one point—Nature is power, full of possibilities and mysteries. But most of the religions of the world except the heterodox ones believe that Nature is not the ultimate power and that God is the supreme reality who controls the Nature. Those who believe in God are called 'theists' and those who don't are called 'atheists'. Between these two belief systems are those who believe that Nature is God and those who believe that it is impossible to find out, prove or disprove whether God exists or not. Both the views are chiefly held by modern philosophers and scientists. Those who believe that it is impossible to know the existence or non-existence of God are called 'agnostic'; the term was coined by the 19th century famous biologist, an anatomist, Thomas Henry Huxley (4 May 1825-29 June 1895).

Whether one finds God or not, he certainly accepts that there is Nature. Nobody can refute that Nature exists. Are we Nature? Are we the parts of Nature or an individual is someone or something beyond it? We have again fallen in a necessity to ascertain what we understand by Nature. Let us recall, we knew about Charles Darwin and his revolutionary concept of Natural selection in chapter 5 where we learnt how Nature works on and within a living being. Darwin puts in his book *On the Origin of Species by Means of Natural Selection, or the Preservation of Favoured Races in the Struggle for Life* the essence of what we today understand by Nature.

". . . . *It has been said that I speak of natural selection as an active power or Deity; but who objects to an author speaking of the attraction of gravity as ruling the movements of the planets? Everyone knows what is meant and is implied by such metaphorical expressions; and they are almost necessary for brevity. So again it is difficult to avoid personifying the word Nature; but I mean by Nature, only the aggregate action and product of many natural laws, and by laws the sequence of events as ascertained by us. With a little familiarity such superficial objections will be forgotten".*

The above passage is one of my favourites; it not only provides a plausible answer to our question but also defines a serious aspect of metaphorical expressions. It is a 19th century passage and Darwin is explaining his fresh concept of Natural selection. The 21st century has a great wealth of systematic findings in this direction; great efforts made in the 20th century; efforts of philosophers, of neuroscientists, of linguists. Metaphor has been central in the development of religion, literary and artistic expressions. This is how 'time' meets with 'gold' when we say 'time is gold'; two different things connected by the essence 'precious'; gold is precious, time is precious, and thus it becomes acceptable to us that time is gold.

The question "Are we Nature?" seems to be important and we may hope that an appropriate answer to this question will improve our understanding of Nature. We learnt in chapter 5 that Nature is full of conditions with incessantly acting forces and laws. Let us visit top-down scientifically with a 'reductionist approach'; who are we? We are living organisms made up of organs; highly regulated. How are they regulated? Everything functions in a certain way because of the inherent structural and functional properties of its elements and under the effect of acting conditions. Organs are made up of tissues and the tissues develop from cells which are the fundamental units of life. We see a world of metabolic pathways inside a cell; biology and chemistry working hand in hand; chemistry being the next level towards bottom. The cells may be complex or simple, made up of large and small molecules; Proteins, Nucleic acids, Carbohydrates, Fats, Vitamins and many other simpler compounds. Molecules are made up of atoms which have subatomic particles, and we see physics operating at the base of everything; particles and waves, mass and energy, matter and anti-matter, dimensions, universal constants with their fixed values, and working principles that always produce the same results. From a scientific viewpoint, therefore, an individual is the product of Nature, part of it, and one may conclude that life is the expression of Nature, consciousness, thoughts and volition the elements of Nature being expressed in an intelligent organism, and in a qualitative sense, everything including a living organism is nothing other than Nature.

But there are few different philosophical and the religious views and beliefs on our question and one cannot ignore them as they are very serious and important. Since there is enough variation in particular beliefs held by these religious and philosophical schools, for the sake of current discussion, it would be better to touch the essence of these views and beliefs later. Chiefly, the most important point against the above reductionist progression is that it leaves 'explanatory gaps'. These gaps arise among different levels throughout the hierarchy of organization. The problem is that these gaps seem to be inexplicable by the same laws that explain the individual levels. Consequently, the men of reason tend to believe that even though fundamentally Nature favours universality of certain laws, there are specific laws and principles operating at particular levels of organization. Biologically, the most of this difficulty lies in explaining some evolutionary aspects, consciousness, complex mental activity and behaviour.

Well, we move towards another point related with Nature and another problem related with modern scientific belief. It is not difficult to a scientist to explain how RNA is formed from DNA and DNA becomes protein in a cell, or how a cell reproduces or regulates itself until it dies. But as he happens to meet a knotty question "what made a molecule biological and different life events set to happen one after another in a series?" and if he does not want to hide behind a vague one word answer like 'time' or 'conditions', he needs to explore what he finally understands by the notions of biological, chemical, physical, metaphysical, cause, effect and chance and so and so on. Similarly, it seems impossible to explain everything with hundred percent precision about an organism, or even about a particle in space. Discoveries after discoveries, inventions after inventions, and an endless thread running through gems after gems, yet something is predictable while something remains unpredictable and it will remain so until a complete and replicable knowledge of Nature is achieved. Whatever science is employed, if our universe is not a tale of creation and the creator, and if there is no logic in waiting for an infinite light-year old witness to know everything about the big-bang and the history of birth of time, space and causality, the role of something that we call as *chance* appears as a naturally undeniable element of Nature.

Thus, while moving towards the explanations of Natural phenomena scientific belief faces two major problems: the problem of explanatory gaps and the problem of limitations of viewpoints. In addition to the scientific beliefs, almost all the philosophical beliefs also face these two problems. On the other side of these problems is the widely recognized notion of chance. By many of us chance is seen as the most unclear element of Nature while by many others, a self-evident prerequisite for each and every event to occur. But as a basic requirement of our enquiry we need to have a clear statement of what we understand by the word chance. I believe that to most of us chance is *the indeterminable, accidental element which can be inferred upon in terms of possibilities by observing a particular set of circumstances.*

If chance is taken as an inherent element of Nature, we have to accept that until now we have travelled on a small arc of a circle, the diameter of which is unknown. But how do we, with both the viewpoints; general and special, arrive at the notion of chance? Is universe an expression of uncertainty or is it completely deterministic? Oh! where the flexibility of our thoughts has brought us, our grip and hanging on the thoughts have suddenly landed us on the surface of age-old notions of 'free-will', 'determinism', 'chance', 'destiny', the surface which gives us a sight of sustainable descriptions of Nature in terms of theories like theory of evolution by natural selection, gravity, electromagnetism, theory of relativity, wave-particle duality and various interpretations of quantum theory, taking our attention towards more recent theories like multiverse theory, string theories and numerous efforts of unifying different theories in a theory of everything (TOE).

With a general viewpoint of orientation towards logic, except the champions of the doctrine of destiny which is regarded as the religious or metaphysical interpretation of determinism and is different from the scientific determinism, almost each of us realizes that life is full of uncertainty, who knows where the journey begins and ends when knowledge itself is a midway event of a passage? Our almost similar everyday experiences tend us to conclude that whether it is a highly dynamic mind or a lump of tissues or combinations of elements, limitations and discontinuities seem to be an inevitable part of everything. To a general viewpoint this is the root to believe in the notion of chance.

With a special, scientific viewpoint, uncertainty is a well-recognized principle. But contrary to artful interpretations by some of us who are not of scientific orientation, the uncertainty principle is not a blindly usable product of fashion-fuelled logic. The principle, published in 1927 by Nobel Laureate physicist Werner Heisenberg (5 December 1901-1 February 1976) is regarded as a fundamental feature of our universe. The principle answers to the question "Is it possible to determine simultaneously the position and the velocity of a particle with absolute exactness?" but to connect with this question one needs to know why to measure the position and the velocity of a particle at a time. Because an accurate measurement would lead us to determine the future and past position and velocity of the particle at any time, further, the accurate measurement of the positions and velocities of each and every particle in the universe would imply the absolute determinism. The principle says that it is impossible to determine simultaneously the position and the momentum (*mass* multiplied by *velocity*) of a particle with absolute accuracy.

What prevents from determining both the things simultaneously with precision? To know the position and velocity of a particle one needs to fall light on the particle. The particle will scatter light falling upon it, and by measuring the scattering one can know its position and the velocity. But there is a limitation; to accurately measure the position of the particle, light of shortest possible wavelength is required, but the shorter the wavelength, the higher the energy of it, and the collision will bring an unpredictable change in the velocity of particle, thus with this short wavelength only the position of the particle could be accurately measured at a given time. Moreover, the light of longer wavelength will accurately measure the velocity of the particle but not the position of it at the same time.

What this limitation shows is that in order to measure the position and velocity of a particle the observer needs to interact with it and this interaction produces a change that limits the accurate determination of one of its conjugate properties. This interaction leaves the observer no longer a passive observer; his observation brings an unpredictable change; it marks a departure from objective to subjective reality. This interpretation is a famous philosophical generalization of quantum theory. The theory has lead to the development of several interpretations after the standard Copenhagen interpretation that was developed in the third decade of 20th century. According to the Copenhagen interpretation the Nature is probabilistic, or in a general sense, the reality can be best described in terms of probabilities.

I must not pretend to be a physicist, I need no courage that unjustly proves me as a master-reader of the book of physics wherein each principle is a huge treatise, but in search of the answers to my loving questions I crept to understand these principles as long as possible knowing unmistakably that perhaps I am the poorest student of mathematics, however, I can perform 5 serious mathematical operations without the help of calculator, they are; addition, subtraction, division, multiplication and percent calculation. I could understand the principles of physics in the only way I knew; the way I understand the fundamentals of life.

Knowing is a possessive affair, and this is why the known is often taken to be true, the clearer the knowledge, the greater the possession with it. We have touched upon uncertainty and the probabilistic view of Nature, now we need to know what science finally tells us about the origin of universe and the origin and flow of time, then we will turn to biological systems; this will give us a clearer view of how far the notion of chance reaches.

We have learnt about uncertainty and probability by looking at experiments with a particle, the particle was an electron; we knew about the principle discovered at the atomic scale. Now, to know about the origin of universe and the origin and flow of time, we are going to jump at cosmology. Certainly, the present discussion requires this jump, yet, a student like me needs much courage to make it worthwhile. Fortunately, I possess this courage which is derived chiefly from three books namely *The First Three Minutes: A Modern View of the Origin of the Universe (1976; Basic books) written by Nobel Laureate*

Deepak Dwivedi

Steven Weinberg, A brief history of time—From the big bang to black holes (1987; Transworld publishers) written by celebrated author and preeminent physicist Stephen Hawking, and The end of certainty: Time, Chaos, and the New Laws of Nature (1997; Odile Jacob) written by Nobel Laureate Ilya Prigogine in collaboration with Isabelle Stengers.

With their bright titles and illuminating contents the above mentioned books narrate the specific concerns of their authors; it will be useful to us to know about their individual ways and certain concerns. As is explicit in Weinberg's other articles and books, he is admirably careful in his descriptions throughout this book and if a reader is fascinated by this trait of the author, he will enjoy more knowing about additional but inseparable 46 seconds after first three minutes of early universe in the chapter entitled *The First Three Minutes* (Weinberg writes ... *the reader will have to forgive my inaccuracy in calling this book The First Three Minutes. It sounded better than The First Three and Three-quarter Minutes*).

In the chapter *First one-hundredth second* he writes about time:

> *"However, although we do not know that it is true, it is at least logically possible that there was a beginning, and that time itself has no meaning before that moment. We are all used to the idea of an absolute zero of temperature. It is impossible to cool anything below -273.16° C, not because it is too hard or because no one has thought of a sufficiently clever refrigerator, but because temperatures lower than absolute zero just*

have no meaning—we cannot have less heat
than no heat at all. In the same way, we may
have to get used to the idea of an absolute zero
of time—a moment in the past beyond which
it is in principle impossible to trace any chain
of cause and effect. The question is open, and
may always remain open".

The scientific knowledge about the origin of universe and time is a golden feat of mankind. In an effort to summarize this knowledge in few lines, realizing that the limitation of the context of this chapter does not leave me free anymore and I may perhaps not be able to write about great works and the scientists who enriched the history of physics with their life-long efforts, I passed weeks in puzzle, added few words a day to this discussion, and you can see that I, in order to hide from my embarrassment grown out of abstractions, have finally stretched a shabby shield.

Astronomy, as a branch of physics will always miss Edwin Hubble (20 November, 1889-28 September, 1953). The path to the verifiable knowledge of origin of Universe includes the knowledge of the earliest subatomic particles, initial events and physical laws operating at its early states, the experiments and groundbreaking discoveries like Doppler Effect, red shift, cosmic microwave background radiation experiments, homogeneity and the cosmological principle, explanations on the basis of general relativity and quantum theory and a search for a new unified theory.

The big bang is presently considered as the standard model of the origin of Universe. Hubble in 1929 noticed that the universe is expanding; galaxies are moving away from us, clearly reflecting that the objects would have been close together in the past. Later it was shown that about ten or twenty thousand million years ago, all the objects were at the same place; this is known as the time of big bang. The earliest Universe was very hot and cooled rapidly during expansion, at this large scale the present Universe is homogenous with little irregularities. Currently we have no justifiable explanation of the earliest fractions of first second, however, there is an open possibility of a theory that could unify the current theories; any such theory, if successfully explains the events at microscopic as well as macroscopic scale, the events related with the expansion of the Universe and creation of matter, will be an unprecedented achievement of human knowledge.

Einstein's work resulted in the unification of space and time, before him the two were considered as separate; now (in 1905 when Einstein published his theory of relativity; the theory had two versions: special relativity and general relativity. General relativity combines relativity with gravity and got its final shape in 1915) they were accepted as a single entity—space-time. Events are points in space-time which is curved. Einstein showed that time is not absolute or a universally constant entity, it changes according to relative motion of observers, and dilates under the effects of relative velocities of observers and gravity. He also predicted that the speed of light is the same value relative to all observers and that nothing can travel faster than the speed of light (186,282 miles per second, in vacuum).

Hawking's narration is one and perhaps the only of its kind and perhaps this is the reason why *A brief history of time—From the big bang to black holes* is one of the most famous science books. The Scientist in Hawking asserts boldly, affirms quickly. This book reflects the brightness of his thoughts especially on the origin and fate of the Universe, origin and flow of time, and his views on anthropic principle. Discussing the singularities and the concept of imaginary time (imaginary time must not be confused with an impression of unreal time, instead it is a mathematical and logical concept of time different from the way we experience it) he writes in the chapter *The origin and fate of the Universe*:

> ". . . . In real time, the universe has a beginning and an end at singularities that form a boundary to space-time and at which the laws of science break down. But in imaginary time, there are no singularities or boundaries . . . . ".

Further, in the chapter *The arrow of time* he elaborates beautifully on different arrows of time:

> "The increase of disorder or entropy with time is one example of what is called an arrow of time, something that distinguishes the past from the future, giving a direction to time. There are at least three different arrows of time. First, there is the thermodynamic arrow of time, the direction of time in which disorder or entropy increases. Then, there is the psychological arrow of time. This is the direction, in which we feel time passes, the direction in which we remember the past but not the future. Finally, there is the cosmological arrow of time. This is the

221

> *direction of time in which the Universe is expanding rather than contracting".*

Prigogine's *The end of certainty* is a unique expression of endeavours of human intellectuals. Apart from Prigogine's great efforts, one thing that deeply influenced me is what appears early in the acknowledgement page of this book. Quoting him, I will then quote famous and influential evolutionist Richard Dawkins from his famous book *The selfish gene (from Preface to second ed. 1989)*. This quoting, for few next lines, will divert us from our current discussion, but will give us another important and timely discussion on an important issue of not only my life but of many of us.

Prigogine writes:

> *"Isabelle Stengers has asked not to be designated as a co-author of this new presentation, but only as my collaborator. Although I felt obliged to respect her wishes, I would like to stress that without her, this book would never have been written. I am most grateful for her assistance".*

Dawkins writes:

> *"I recently learned a disagreeable fact: there are influential scientists in the habit of putting their names to publications in whose composition they have played no part. Apparently some senior scientists claim joint authorship of a paper when all that they have contributed is bench space, grant money and*

*an editorial read-through of the manuscript.
For all I know, entire scientific reputations
may have been built on the work of students
and colleagues! I don't know what can be done
to combat this dishonesty . . . ."*

*(By permission of Oxford University Press; The selfish
gene 3E by Richard Dawkins, 2006)*

If you have a scientific background, the former will touch you with the great concerns of the authors (*Isabelle Stengers* and Prigogine), but the latter will streak your head with a bitter reality. The former is rare; the latter is prevalent in current scientific atmosphere. It is 2013; nothing seems to be changing except the increasing severity and victimizations. My salutations to Isabelle Stengers, Ilya Prigogine and Richard Dawkins! Again, and once again. May we hope that the above expressions of these intellectuals will prove their importance being more than just an expression of personal feelings?

During some critical phases of my most recent academic struggles, often I heard from other victims "We should have a hidden camera with us while interacting with these wild powerful people". Well, I could have a hidden camera or simply my mobile voice recorder switched on, and then? Suppose I grunted at them showing the evidences of their mistakes, would then those wild powerful people understand others better? Never, because they have gradually grown ignorant of moral values, however they would be afraid of me. Then? Suppose they anyhow managed to blackmail me and then a long chain of tricks and threats and warnings, ultimately up to the effect of their imprisonment. Yet, I would feel bad, why did it happen? What would be the health of a morality that cannot grow without fear? Are humans so apt to grow wild? Often a strange thought troubles me "Are moral values rapidly losing their charm? Must we explore special paths to reach them to the heart part of minds? Or otherwise be ready to step into an advanced future where morality is nothing more than an interesting monument of rudimentary human brains?"

I think that the above discussion warrants its importance not only in education and research but also in other joint enterprises. Let us not grow oblivious of time because we knew earlier how time becomes gold. Interestingly, looking at a few points from Prigogine that he preserved in *The end of certainty* and my position on the notion of chance, we will turn towards the religious-philosophical viewpoints on the origin of Universe and the origin and flow of time. Prigogine's views differ from Hawking's on several aspects, on the question of beginning and end of time, he writes:

> *"Briefly stated, however, we believe that the big bang was an event associated with an instability within the medium that produced our universe. It marked the start of our universe but not the start of time. Although our universe has an age, the medium that produced our universe has none. Time has no beginning, and probably no end. But here we enter the world of speculation. The main purpose of this book is to present the formulation of the laws of nature within the range of low energies. This is the domain of macroscopic physics, chemistry, and biology.*

*It is the domain in which human existence
actually takes place".*

Prigogine defines the two grand expressions of Nature with the viewpoint of indeterminism suggested by understanding the properties of instable systems and irreversibility:

> *"Today we are not afraid of the "indeterministic hypothesis." It is the natural outcome of the modern theory of instability and chaos. Once we have an arrow of time, we understand immediately the two main characteristics of nature: its unity and its diversity: unity, because the arrow of time is common to all parts of the universe (your future is my future; the future of the sun is the future of any other star); diversity, as in the room where I write, because there is air, a mixture of gases that has more or less reached thermal equilibrium and is in a state of molecular disorder, and there are the beautiful flowers arranged by my wife, which are objects far from equilibrium, highly organized thanks to temporal, irreversible, nonequilibrium processes. No formulation of the laws of nature that does not take into account this constructive role of time can ever be satisfactory".*

As most biologists do, I take the notion of chance very seriously. The reason is clear; a dynamic biological system is in itself a world of interacting structures and pathways, highly integrated and regulated, varying under balanced and steady states. Variations produced by mutation and genetic recombination in turn produce different phenotypes; structural and functional, up to the level of observable behaviour of the organism. Small variations at different levels of structural and functional hierarchy are able to produce big changes as is explained by *butterfly effect* of chaos theory. All this takes us to the variations with which molecules interact with one another, and despite the fact that each living being has an individual life, the new vistas for this life are always open because life depends on *interactions,* we learn this when we study 'complex systems' with different viewpoints, an example of which is a recently growing approach known as 'systems biology' which says that *the whole is greater than the sum of the parts* or in the words of P. W. Anderson *"more is different"* (*more is different* was published in 1972 in *Science*). Evolution reflects the importance of time; clearly, it is time and chance dependent, and in whatever way we define chance and time life is never separate from them.

From this point we get close to very well known and widely discussed duo—'determinism' and 'free will'. The above discussion on the notion of chance has already conveyed us what is meant by determinism. There is no doubt that these notions are important, but what makes them so important? They are important because they have serious impacts on the social values, scientific—philosophical quest and personal belief. We have seen that the notion of hard determinism is not sustainable because there is a role of chance in the occurrence of events, similarly, absolute free will is not sustainable; no free will can exist in complete isolation from the ground which gives rise to it, whatever name we give to this ground—an individual or brain or mind or stimulus or cause or Nature or something else.

The above notions have a range of issues associated with them; spanning from moral principles to law. Law improves the shape of societies, it makes them symmetrical. But free will may seem pushing law into the witness stand: why should a culprit be punished if his will was not completely free from the ground which tempted or compelled him to do something punishable? But we see that there is enough space to think that the culprit could have done something else; he could have chosen another option to go through in order to prevent himself from being a culprit. We see that *the availability of options* and *the ability to make choices* becomes a necessary condition for defining free will. Let us swiftly turn towards 'behaviourism' to focus on a central aspect of the human and animal behaviour. I hope the following free verse which I composed under the title *I wish they knew* will provide an economical and clear cut view of the commonest aspect of behaviour.

Bees
Such a house they build!
Sweet, hexagonal, aromatic
I wish they knew
What a beauty they own!
How brilliant their work!
How lovely their dance!
In this evolutionary world
Of being and behaving
Very few have come to know
Who they are
And what they are doing

Is a dog guilty if bit a passerby? The justification lies in the ground which gives rise to the freedom of will. Dog cannot contemplate, it cannot *think* of options; yet it possesses certain degree of awareness with which it is able to make a choice. Think of two tame birds: one inside a cage and the other domesticated outside the cage. Which one looks freer? Is a bird flying freely in a forest actually free? If yes then go on; is life free from what it grows out of or expresses through or can sustain in? No, because like every natural phenomenon, life requires a ground, a set of conditions, or in a loose sense, a medium, without this ground or medium it cannot exist. Then how could a will become independent of everything if it has to exist in exactly the same set of conditions? There must be something that determines the course of 'attention' to the effect of expression of will.

The above assertion compels us to have a different view of free will. There is no dearth of ways and views, philosophy and psychology have marvelled in exploring the ways; here is another one: *free will is a subjective expression;* someone has reasons to say "I have free will" and so does the bird flying freely in the forest: he may like to explain it with *the frame of references* and *relativistic empiricism.* But he may also ask you in reaction "do you know what is subjective and objective and phenomenology? Do you know *what it is like to be a bat?*" ('What it is like to be a bat' is a famous and influential paper published by Thomas Nagel in 1974).

But he has to admit that the things other than 'freedom' are also compatible and often attached with 'will' for example, 'available options' (objective) and 'individual's selectivity' (subjective). Suddenly 'free will' sounds as an exaggeration; only 'will' with 'available options' and 'individual's selectivity' appears to be sustainable with both the subjective and objective viewpoints. But when we try to dissolve the boundaries of subjective and objective in order to keep free will as a whole, intact, we often construct a puzzle stressing on the compatibility of 'free' and 'will' and then in attempting to unify free will and determinism; such efforts succeed in saying nothing more than a superimposition of an 'action' on 'the process'.

Though there are certain limitations of reductionism, its contribution to present human knowledge is immense. The same is true about behaviourism and developmental psychology. But a recent tendency has caught much criticism: many scientists try to explain animal behaviour by emphasizing the molecular and cellular events; they try to explain the characteristic features of an animal population by gene interactions and expressions but the genotype-phenotype scheme has yet wider prospects some of which can be noticed in the recent definitions of phenotype extended to the behavioural characteristics of an organism. Genetics, genomics, proteomics, molecular biology, cell biology, neuroscience and other specialized branches of biology have proven their unparalleled importance but is it appropriate to define each and every aspect of an individual's life by emphasizing on *genes* and *neurons*? Is a life that has a culture so simple to define and predict? Certainly not, and this is why there are more recent approaches like modular biology and systems biology.

Questions are tremendous catalysts. An interesting question is now visible: Is behaviour shaped by Inheritance or by experience? This question can be made clearer with the help of two questions: what makes each and every animal to behave in the same way as its parents did?

Then what makes animals to behave slightly differently in different environments?

These questions clearly concern the science of behaviour. they take us to the great works of J B Watson (9 January, 1878-25 September, 1958), the founder of behaviourist school, Influential ethologists and ornithologists Konrad Lorenz (7 November, 1903-27 February, 1989), Karl von Frisch (20 November, 1886-12 June, 1982) and Nikolaas Tinbergen (15 April 1907-21 December 1988) who shared 1973 Nobel prize in Physiology or Medicine, B F Skinner (20 March, 1904-18 August, 1990) who contributed the science and philosophy with his prolific works and the evolutionary biologist John Maynard Smith (6 January, 1920-19 April, 2004) who is famous for his two major contributions; Game theory and Evolutionary Stable Strategy (ESS).

The present context gives me an opportunity to discuss a brilliant observation made by few scientists. I knew about it from my favourite book *Life-The science of biology* written by William K. Purves, David E. Sadava, Gordon H. Orians and H. Craig Heller The influence of this book on me can also be noticed from chapter-5 where I mentioned it twice in the same paragraph. Sadava and Heller begin chapter 52 *Animal behavior* with the important observation made by the scientists:

> *"A troop of Japanese macaques living on an island was being studied by scientists, who fed the monkeys by throwing pieces of sweet potatoes onto the beach from a passing boat. The monkeys tried to brush the sand off the sweet potatoes, but they were still gritty. One day a young female monkey began taking her sweet potatoes to the water and washing them.*

Soon her siblings and other juveniles in her play group imitated her new behavior. Next their mothers began washing their potatoes. No adult males imitated this behavior, but young males learned the behavior from their mothers and their siblings. The scientists were fascinated by the way the creative, insightful behavior of one juvenile female spread through the population, so they presented the monkeys with a new challenge: They threw wheat onto the beach. Picking grains of wheat out of the sand was tedious and difficult. The same juvenile female came up with a solution: She carried handfuls of sand and grain to the water and threw them in. The sand sank but the grain floated, enabling her to skim it off the surface and eat it. This behavior spread through the population in the same way potato washing did—first to other juveniles, then to mothers, and then from mothers to both their male and female offspring."

*"The macaques now routinely wash their food. They play in the water, which they did not do before, and they have added some marine items to their diet. Clearly, this population of monkeys has invented new behaviors that have spread by imitative learning and have become traditions in the population. One could say that they have acquired a **culture**: a set of behaviours shared by members of the population and transmitted by learned traditions."*

(By permission of SINAUER ASSOCIATES Inc. Sunderland MA)

The above story explains the key features of behaviour: *imitation* and *emulation*. The story on a careful examination reveals that behaviour may vary significantly under the effects of environment and experience though it largely depends on inheritance. The experiments conducted by the above mentioned behaviourists and evolutionary biologists including others have confirmed that this conclusion is relevant. In humans, behaviour varies at much higher degrees.

But does an animal understand what it is doing? Or it just does under the effects of its inherent instincts and reflexes? Studies on complex behaviours including learned behaviours show that different animals have different degrees of awareness yet these degrees are not sufficient to make them think because they lack required organization of brain for this purpose, therefore they are not aware of what they are doing but they are aware of dangers, food and their needs like hunger, thirst and other members around them etc. and so they behave in a certain way. This type of awareness does not need thinking. The effects of organization and development of brain with respect to behaviour can also be observed at initial stages of human development, this science is known as developmental psychology. Developmental psychologist Jean Piaget (9 August 1896-17 September 1980) proposed the *theory of cognitive development* according to which humans have four development stages, as a child grows, egocentrism weakens and logical thinking begins to develop, abstract reasoning develops in later stages of childhood.

Though life can thrive without brain and volition, but in that case it is governed almost completely by the environment. Brain is necessary for subjective world, for self-awareness which is an attribute of it and gives it the comprehension of its own competence, and this comprehension adds another dimension to the better survival of the individual. The presence of brain in an animal does not mean that it is self-aware or it can think. The brain activity depends on its structural and functional organization, the more complex the brain the higher functions it can perform.

Human brain possesses unique structural and functional characteristics. I knew this fact by knowing about two things: the great works of neuroscientists and philosophers, and my life. You might have wondered at my repeated assertions regarding my foolishness. It is so because I love this fool in me; it is this fool who prompts me to walk into different lands and provides me the reasons for why I must stay in these lands. Each and every beginning of my thoughtful life meets this fool first. I was really fortunate that I could come face-to-face with this fool. I have witnessed this fool's aimlessness, fears, restlessness and tears. He held me during my wondering, during my dizziness, he consoled me when people ridiculed at me on my seemingly futile but sincere efforts. Undoubtedly I am I am I am for most of mine this fool.

Emotions and feelings affect reason. This is an important conclusion by famous neuroscientist Antonio Damasio. Neuroscience is the major contributor to our knowledge of brain. Damasio explains in his book *Descartes' Error: Emotion, Reason, and the Human Brain* (Putnam, 1994) that emotions are involved in reasoning. He begins chapter-4 entitled *In colder blood* with a popular belief:

> *"There never has been any doubt that, under certain circumstances, emotion disrupts reasoning. The evidence is abundant and constitutes the source for the sound advice with which we have been brought up. Keep a*

> *cool head, hold emotions at bay! Do not let your passions interfere with your judgment. As a result, we usually conceive of emotion as a supernumerary mental faculty, an unsolicited, nature-ordained accompaniment to our rational thinking. If emotion is pleasurable, we enjoy it as a luxury; if it is painful, we suffer it as an unwelcome intrusion. In either case, the sage will advise us, we should experience emotion and feeling in only judicious amounts. We should be reasonable."*

Further he says in chapter-6 entitled *Biological regulation and survival*:

> *"Emotions and feelings, which are central to the view of rationality I am proposing, are a powerful manifestation of drives and instincts, part and parcel of their workings."*

("In Colder Blood", "Biological Regulation and Survival", from DESCARTES' ERROR by Antonio R. Damasio, MD, copyright (c) 1994 by Antonio R. Damasio, MD. Used by permission of G. P. Putnam's Sons, a division of Penguin Group (USA) LLC.)

Neuroscience has now become a big area of scientific pursuits; thousands of scientists are presently working on different problems related with human brain and mind. To me, it is necessary and worth mentioning about the organizations and groups working for spreading ideas and knowledge. In 2011, during my haphazard searching of literature on internet, I came to know about

TED (Technology, Entertainment, Design) talks and conferences, I went through some of the freely available talks. These talks gave me a new way of knowing, later I found several other educational videos on You Tube. Google is now a synonym for internet searching and a student's inseparable companion, my sincere thanks to each of these charioteers of knowledge.

The neuroscience of emotion, feeling, reason, belief, experience and behaviour has established numerous facts and has given us important insights into the human nature. There are scientists who have studied the neurological problems inside out; that is they have investigated their own neurological disorders. It is very challenging to someone to apply logic and reason efficiently while passing through an emotional and mentally distorting phase of life. Famous and influential scientists and authors Jill Bolte Taylor and Oliver Sacks are the two present examples of such a challenging endeavour.

Jill's *My stroke of insight* is a top watched TED talk and her book *My Stroke of Insight: A Brain Scientist's Personal Journey* a famous book.

Jill's insights are admirable, however, the world of wisdom can never lie away from the loops of criticism. Love is real, but reality has definable limits and these limits may prompt us to question how far is love real? Answers seem to be begging for vigour. I believe that the experience and the explanation of experience are profoundly different things. It is quite like the problem of *Qualia* or *self* or act of *mirror neurons* and is about representations of representations in the brain. To have an experience is like walking through a narrow aisle, experience is about the biological economy and regulation; it is a tenant of evolution. Then, experience is not just a passage, processing or generation of information, it is also an optimization working on system level.

Oliver Sacks is a celebrated author and neuroscientist. He is well known for his studies on patients with neurological problems. I came to know about his works on hallucinations when I saw his famous TED Talk *What hallucination reveals about our minds* in which he speaks about his own hallucinations (geometric hallucinations) after discussing a case of visual hallucinations. He begins his talk with the following words:

> *"We see with the eyes, but we see with the brain as well. And seeing with the brain is often called imagination. And we are familiar with the landscapes of our own imagination, our inscapes. We've lived with them all our lives. But there are also hallucinations as well, and hallucinations are completely different. They don't seem to be of our creation. They don't seem to be under our control. They seem to come from the outside, and to mimic perception."*

Then he discusses about the case; an old lady in her 90s who was perfectly sane, lucid and of good intelligence, and was blind for five years. But now, for the last few days, she'd been seeing things. She said, *"People in Eastern dress, in drapes, walking up and down stairs. A man who turns towards me and smiles. But he has huge teeth on one side of his mouth. Animals too. I see a white building. It's snowing, a soft snow. I see this horse with a harness, dragging the snow away. Then, one night, the scene changes. I see cats and dogs walking towards me. They come to a certain point and then stop. Then it changes again. I see a lot of children. They are walking up and down stairs. They wear bright colors, rose and blue, like Eastern dress."*

Further Oliver says, *"Well, I examined her carefully. She was a bright old lady, perfectly sane. She had no medical problems. She wasn't on any medications which could produce hallucinations. But she was blind. And I then said to her, "I think I know what you have." I said, "There is a special form of visual hallucination which may go with deteriorating vision or blindness. This was originally described," I said, "right back in the 18th century, by a man called Charles Bonnet. And you have Charles Bonnet syndrome. There is nothing wrong with your brain. There is nothing wrong with your mind. You have Charles Bonnet syndrome."*

Oliver explains, *"Now this, for me, is a common situation. I work in old-age homes, largely. I see a lot of elderly people who are hearing impaired or visually impaired. About 10 percent of the hearing impaired people get musical hallucinations. And about 10 percent of the visually impaired people get visual hallucinations. You don't have to be completely blind, only sufficiently impaired."* Then he explains briefly how the structural and functional impairments in different brain areas along with the rational activity of brain may give rise to different kinds of hallucinations.

This important talk made me very serious about my grandmother's situation, however, she was no longer alive by this time. Recall from chapter-2, for more than a decade in my juvenile days I had been trying to understand her world, to find the basis of her experiences that few people regarded as *divine* while few others, *crazy*. My awareness towards hallucinations draped me in an utterly critical thought "Was my grandmother hallucinated?" *hallucination* crammed my emotions in a narrow bracket; a bracket made of twisted boundaries.

Two types of special viewpoints, like always, are now eyeball to eyeball. Both have ample reasons to claim that they are special: one is the view of science; the other is the view of religion. Philosophical insights; analyses, conclusions and inferences are apparent in both the views and this is why many scientific and religious views are often philosophical not only in outlook but in the method as well. Languages, metaphorical and poetic expressions are equally important in this regard because they concern the human brain as directly and deeply as the above special viewpoints.

Though there are always different branches and schools of thoughts in each and every viewpoint yet the scientific viewpoint is generally considered as the most robust view of life. This is so because science works with testable predictions and can explain the individual effects in relation with their individual causes. Reproducible evidence is the heart of science.

Religion on the other hand implicates common and personal beliefs. No religion in the world refutes the role of human comprehension and competence in the construction of belief, moreover, each and every religious belief is chiefly concerned with the social, individual and soteriological (salvation, healing and liberation) achievements of any individual life. Thus no one can deny that science and religion both chiefly concern observable and explainable events of Nature and that they both are concerned with the better way of living, the only difference is the way they see the life through.

But religion concerns something more. This *more* is often something that transcends what we generally call as 'real'. *Mysticism* is no longer welcomed in science and philosophy (except in metaphysics), but it is preserved in religion. Scientific view says, "Nothing is mysterious, there is no mystery-stuff in a real sense" while religious view says, "Mysteries will continue to flash because human knowledge is a long tale of limits, man is not the lord of Universe, he is not even the lord of life". One can understand that it is the song of lord which holds special position in religion.

'Lord' seems inexplicable through strict scientific viewpoint, while through religious viewpoint 'the explanation of lord' is a deviation from 'song of lord', and this is very important because *devotion*, the heart of religion, has nothing to do with explanation but peaks unconditionally on song strings of love. Science cannot presently afford to explain *devotion* and *loyalty* on its own, it has to travel long to reach even on a sustainable explanation of conscious experience and representations, physiology of different kinds and states of love. Religion, on the question of explanation concerns chiefly the explanation of virtues and ideal path to lives related with a life.

But why do we see so much extremities and distorted forms of religious beliefs around us? Because we often confuse religion with something else, something soulless and heartless that mimics in traces the essence of religious belief. I am speaking this on the basis of my struggles into this direction, which, on few occasions were within an inch of the question of life and death, I briefly mentioned about some of those experiences in chapter-4, 5 and 6; Imagine of a religious place where you have to wander for this long with all your serious efforts to understand the essence of religion! Think of the bewilderment arose due the differences between numerous commentaries on a pure, buried thought!

Religious explanations on the *cause* of Universe and life especially the human life differ widely. Differences are never a problem, they are in fact necessary for the sake of reason, for a meaningful criticism, but the problem is that these explanations *vary in essence*, and they vary much; they vary on the question of ultimate reality, they vary on the question of causes and effects, religions differ, schools of individual religions differ, individual champions within individual schools differ. It is pertinent to quote physicist David Deutsch; from his TED Talk *A new way to explain explanation* that a bad explanation is easy to vary, a good explanation on the contrary, is hard to vary.

Each and every religious belief claims to be the truest and the most important knowledge. Isn't it a striving to thrive on mileage of explanation while chiefly depending on but clearly deviating from the 'song of lord aspect'? Isn't it a self-contradiction? Then why should a religious belief often keep away from the evidence based explanation of Nature? Why does this belief hesitate to believe that the limitation of knowledge does not mean that an effort should not be made by the brain? Even if head is prostrated, brain must not fall asleep; it should not shirk its duties too.

Grandmother's point of view and her world of experience will always remain special to me. Decoding her subjective world is now impossible, no explanation can sustain in her perpetual absence. Life seems to be the tenant of time. I often try to paint my emotions; let me paint the emotions with bright colours of science. One thing that I take very seriously is rare occurrence of the psychic experiences that prove to be true, or that often occur with coincidences, I wrote about such experiences of mine in chapter 4, 5 and 6.

Superstition is a challenge before justifiable belief. But belief is a fundamental trait of intelligent and social life and has full rights to take birth even in the lack of evidence. No doubt science has validated a great wealth of belief systems yet much is left for science to deal with different types of beliefs which it often ignores. Being a science student who has observed *something* that cannot be taken casually, I believe that meanings don't exist outside the Nature, that there is no inside-outside of Nature, and that a whole new world is still unexplored, untouched by science. We must not ignore the fact that there are belief systems that are near extinction sunken beneath between reasonable explanation and hypocrisy.

To me it is important to know the bigger truth ever, thus to believe in an open possibility of knowing the ultimate truth of everything. Justification for each and every belief is equally important. Presently I have no way to prove or disprove the nature and existence of God but I believe that depending on the position of the observer, questions become direct or indirect, appear as close or distant and the answers who are their counterparts in the land of logic, can be shaped to complement them better. I believe that no fact of Nature is useless to an observer, I must connect appropriately rather than rapidly, my belief must flourish up on reason, and my intellect must explore different ways. I am not an atheist nor a theist nor even an agnostic, I am that who is aware that someone or something is in question, I believe in solutions, in joining the answers with questions, I believe in continuity of sustainable beliefs, to make the unknown known, and on the question of God, I would like to call myself a *Cognist*.

The word 'cognist' along with my definition of it, can be taken in a slightly different sense, but Cognition is a widely known term. I generally avoid inventing new words, and it worries me much looking at the modern trend of reckless use of fresh, catchy hybrid words. A cognist is he who believes in justifiable knowledge and the processes or way through which knowledge is arrived at. A cognist believes that if God exists, the knowledge or description of God is possible, but if he is not able to decide on the existence of God at a given time, he will not leave out the question of God saying like an agnostic that it is impossible for a man to prove the existence of God. A cognist believes that if a man can find out what light is, how it travels and with what speed, he can also find out whether God exists or not. A cognist is aware of the evidences that an atheist shows him asserting that there is no God but he is not in haste to jump on the conclusion as his atheist friend does, because he cannot ignore the gaps, he knows that enough remains unknown and is to be made known. Same thing can be applied to his position with the positions of his theist friends; a cognist is ready to take up the responsibility of defining what or whom we often confuse with God and what we actually mean by saying that *God exists*.

Gardener
Will a wish be granted?
O Gardener
Grant my wish Gardener

My longing and my hope
Have held each other
In the garden
Grant my wish Gardener

My life waits for a moment
Incessantly
Before it reaches
Grant my wish Gardener

Wind loves you, I know
Embosoms you, I know
Say to bring fragrances
Of dawn
Of innocence
Of growth
Of devotion
Of drops
Of love
Of soil
Of peace
Fragrances are guests, I know
Soft are their ends, I know
Will you join the ends?
With your magic hands
Adorn me O Gardener
With this garland
Grant my wish Gardener

Seasons are curly, they know
Flowers are lovely, they know
Let them pick a thread, they tuck
Colours

Deepak Dwivedi

Sights
Surface
Shelter
Pious beads
Freedom seeds
And touch them once so they know
Someone believes in them
Believe them
They know me not
They know you not
Touch them once O Gardener
Grant my wish Gardener

Buds are lonely, the Garden knows
Tree is sad, the Garden knows
Send bees
With songs
Of way
Of arrival
Of stay
Of sunshine
Of gain
Of nectar
Of rain
The Garden knows
Your hut O Gardener
I know not
And wander
Let me be the Garden O Gardener
Grant
Grant my wish Gardener

I am influenced by great thinkers; highly and deeply. The Earth is a practically great planet. Therefore each and every nation is great by birth yet in a more practical sense, the greatness of nations depends majorly on the greats who make them great.

But this produces discrimination; advanced nation and less developed nation, powerful nation and diseased nation. The pride is natural but on the critical side of a coin, one has to accept that deviating from their individual growth and security, nations almost always possess a spinning selfishness; they develop weapons, and have rivals, politicians and diplomats under the labels of alliance, friendship and peace.

Humanity is great, no doubts, but do all humans want to live like greats? National pride is a great feeling, no doubts, but aren't there nations who look keen on attacking? Look at history without any delay. Know the history of significant losses. If humanity is really in search of a better century than ever, it must dissolve the toxic barriers, poisonous feelings and lethal actions. Enough is there to fight against; fight against diseases, fight against natural calamities, fight against constraints of peaceful life, develop the best shields against diseases and disasters. Fortunately, today we have the nations that never attack yet there is something to worry; too much shields themselves may prove dangerous.

Do you see different states of a nation afraid of their capitals or of national capital? No, because there is no place, no route for fear, and stands only one feeling; oneness. Believe it or not, the feeling of oneness is really different from the feeling of unity. It is not tough to understand why nations feel so insecure. Look at the borders plunged in ploys to grow wider, longer, wilder.

Deepak Dwivedi

My words are breathing a noise; perhaps they are growing critically worried; I am sorry if they hurt.

Nonetheless, I must respect the questions heading towards me. Many of us may raise serious questions in response to the above points "Deepak, what do you know about National policies and management? Are you aware of the constraints and the challenges of the *oneness* that you are arguing for? Do you know the practicality of this *oneness* precisely? Do you know how speculative your pen has made this page?"

Answers bring the questions at peace. Let me attend each of the above questions individually.

Question: "Deepak, what do you know about National policies and management?"

Answer: Though I don't deserve to say that I know nothing about National policies and management, but let us suppose that I know nothing and that collectively we know *something* about them.

Question: "Are you aware of the constraints and the challenges of the *oneness* that you are arguing for?"

Answer: I don't need such awareness. Those who are thoroughly aware of such awareness knowingly or unknowingly construct a noisy cell of limitations. Peace, by its very nature can neither afford uprooting somersaults nor can it become louder than silence.

Question: "Do you know the practicality of this *oneness* precisely?"

Answer: No I don't know; please help me to make it precise. Let us start from "*oneness* is more economical and more organised expression of social life".

Question: "Do you know how speculative your pen has made this page?"

Answer: The momentum and the ink of this pen are real; this page keeps staring at *something*, it has the signature of a hope.

This hope is great, suspicion cannot contaminate it; it has a powerful feeling, a robust emotion and a vivid reason. This subjective experience of *oneness* has enormous positivity. It embraces the attitude with great warmth. Hope attracts. But the *logic of competition* says that the positive attitude is never separate from selfishness. The greater fitness of individuals will bring more competition and competition has both sides; positive and negative. Another trouble is there; sometimes fitness of an organism demands an amputation of the organ that is diseased and untreatable; removal of the dying organ becomes necessary so that the spread of disease can be checked. Think of a family in which a diseased member's survival is now a danger to it. Suddenly the life becomes statistical; significant and non-significant values. And I am talking about global *oneness*! Can *oneness* solve these troubles first?

Yes it can. I accept that the path to achieve *oneness* is long but it is not unknown. *Oneness* itself has miles to go before it meets its grandness. Let all the sciences, all the philosophies, all the branches of human knowledge see each other through the *principle of oneness*, let all the mirrors of neurons receive the rays, let the distance be redefined. I believe that our lives are the parts of a thread. I am a student of *oneness,* and am waiting for my colleagues.